Understanding Physics Using Mathematical Reasoning

Andrzej Sokolowski

Understanding Physics Using Mathematical Reasoning

A Modeling Approach for Practitioners and Researchers

Springer

Andrzej Sokolowski
Division of Mathematics and Science
Lone Star College
Houston, TX, USA

ISBN 978-3-030-80207-3 ISBN 978-3-030-80205-9 (eBook)
https://doi.org/10.1007/978-3-030-80205-9

This Springer imprint is published by the registered company Springer Nature Switzerland AG
The registered company address is: Gewerbestrasse 11, 6330 Cham, Switzerland

Preface

The book is addressed to physics practitioners, researchers, and students who strive to present physics beyond the boundaries of traditional curricula. It aspires to propose instructional approaches that make the physics content exciting to teach, learn, and explore. Situated in contemporary physics education research findings, it encourages the readers to consider mathematical reasoning as a convenient tool to predict, extend, and verify scientific theories. While mathematical reasoning can take various layers of complexity, this book focuses on developing covariational sense and graphing techniques supported often by limiting case analysis. A synthesis of research supported a design of an empirical–mathematical modeling scheme and ready-to-use classroom instructional materials guided by the design. Emphasized correlation to mathematics methods makes the book an appealing resource also for mathematics educators and researchers who are interested in mathematical modeling.

The Book Structure

The book consists of three parts composed of 13 chapters whose scope and sequence will lead the reader to discover many opportunities for extending inquiry in physics by taking advantage of mathematical reasoning tools.

Part I consists of two chapters; it brings forth physics building blocks such as laws, principles, theories, and theorems and reflects on these constructs through the prism of the structural math domain. While these constructs are frequently used in teaching practice, their algebraic entanglements are often overshadowed by their scientific meanings. Research in scientific modeling as an environment nurturing the link of scientific and mathematical reasoning skills is also discussed. Chapter 1 of Part I examines the potential of the laws and principles to be expressed as covariate algebraic structures. While mathematics and physics are different subjects, they share similar philosophies that when realized make students' math knowledge transitioning to physics comfortable. Chapter 2 delves deeper into the methodologies of physics and mathematics, revealing common teaching goals that help to establish pathways to allow the structural domain of math knowledge to transition to the physics classroom.

Part II, consisting of three chapters, is dedicated to establishing links between current recommendations in physics education and how the research in mathematics education can respond to these recommendations. Chapter 3 of Part II summaries recent findings on using mathematical modeling in physics practice. Although the literature provides several modeling cycles, mathematical reasoning is not explicitly underlined; therefore, a necessity to develop a new model emerged. In Chap. 4, empirical–mathematical modeling scheme is proposed, and its phases are described. The scheme provides a general framework of labs and lecture structure designs that extends students' scientific reasoning by considering natural phenomena as covariate relations. Mathematical reasoning has its established place in physics education; however, its more powerful version, covariate reasoning, is still being researched. Chapter 5 provides a more pragmatic view on using covariate reasoning, and in addition, it discusses the main underpinnings of limiting case analysis that serves as a tool to quantify a hypothetical system's behavior. An attempt to arrange physics formulas as classes of various types of covariate entities is also made.

Part III is considered the core of the book. It consists of eight case studies where the theoretical underpinnings outlined in Parts I and II are placed into practice and evaluated. While in Chap. 6 of Part II, limiting case analysis extends the investigation of the motion of a system of two objects, Chap. 7 proposes a reconstruction of Newton's law of universal gravity to include gravitational fields as a mediating parameter nurturing covariate reasoning. In Chap. 8, applications of parametric equations to enhance the understanding of projectile motion as a two-dimensional motion are examined. Chapter 9 investigates the opportunity to generalize image characteristics stemming from the lens equation considered an algebraic function. In the traditional science curriculum, the concept of the mole is associated with the content of the chemistry curriculum; however, it is also a physics concept. An effort to induce math reasoning to understand the unique measure of mole and its conversions is included in Chap. 10. An attempt to reconstruct the Einstein formula for the photoelectric effect that allows reason mathematically is presented in Chap. 11. Chapters 12 and 13 aim differently and examine students' perspectives of applying math structural domain to understand physics. Each case study concludes with suggestions for further explorations.

Houston, TX, USA Andrzej Sokolowski

Contents

Part I
Conceptual Background

Chapter 1
Physics Constructs Viewed Through the Prism of Mathematics

1.1 Mathematics as an Indispensable Part of Physics Inquiry

Among academic disciplines, physics is one of the oldest (Krupp, 2003), and it is defined as a general analysis of nature conducted to understand how nature behaves (Young & Freedman, 2004). Physicists observe phenomena in the pursuit to identify patterns, explain these patterns following scientific inquiry, and apply mathematical apparatus, if plausible, to express the patterns concisely. These patterns can be supported by theories and advance to formulas called laws, principles, or theorems. Physics constructs are often expressed using algebraic symbols and the rules of mathematics. Some physicists claimed that the elegance and logic of physical laws are only apparent when expressed in the appropriate mathematical framework (Schwartz, 2012). According to Feynman (2005), mathematics provides the tools to express the phenomena' dynamics using a universal language. Mathematics appears as the perfect tool in which one can narrate a natural phenomenon (Agnew et al., 2009) because nature is the domain of measure and order that responds only to questions expressed in a mathematical language (Koyré, 1957). While mathematics rules provide a framework for phenomena quantification, phenomena' representations cannot be reduced to constructing mathematical formulas and manipulating their parameters using mathematical algorithms. Mathematics helps model data, formulate the relations of interests in symbolic forms and extend the inquiry into a hypothetical domain. However, what distinguishes these representations in physics from how they are used in a context-free mathematical sense is that the scientific equations are built of quantities with definite physical interpretations reflecting their derivation from nature. Once embraced in algebraic structures, they provide opportunities to apply algorithms to find unique values and reason beyond the boundaries of their measurable magnitudes.

The content of physics has been formulated through investigations conducted by experimental and theoretical physicists. Experimental physicists apply scientific

A. Sokolowski, *Understanding Physics Using Mathematical Reasoning*,
https://doi.org/10.1007/978-3-030-80205-9_1

methods often followed by inductive reasoning to formulate theories, laws, principles, and theorems. They observe the behaviors, identify parameters, select parameters of interest, gather data, and formulate general mathematical models, if plausible. Theoretical physicists apply different scientific methods, but the aim is similar; they make observations, construct theories, formulate mathematical models, but apply a deductive inquiry to determine if their models support the theories. Theoretical physicists also use mathematics as a reasoning tool while applying deductive inquiry. The success of their investigations depends on how well the theoretical results agree with observations of nature. In the case of dissonance, the mathematical representations, not the facts observed, are being reexamined.

The products of investigations by experimental and theoretical physicists are similar. The verbal explanations of these constructs are called theories. The following sections provide more details about each of these fundamental physics constructs.

1.2 Laws of Physics and Their Mathematical Embodiments

Phenomena behavior is typically described by establishing relationships between selected quantities, leading to formulating a law. Physics laws reflect casual relationships within phenomena and are formulated through data taking and mathematical modeling (Wigner, 1990). Laws do not explain the phenomena. Laws describe the phenomena. To be called a law, the report must be conducted and validated over a wide range of observed experiments of a similar nature and aim. Bhushan and Rosenfeld (2000) claimed that the derived laws are the best approximation of nature's more profound behavior that is still waiting to be fully discovered. While the laws are established and widely accepted, according to Bhushan, a room for discoveries or modifications of the existing ones still exists, which can be an inspiring message for the students. Laws as descriptive principles of nature must hold in all circumstances covered by the laws (Oxford University Press, 2015, p.212). Scientists give the title *law* to certain concise but general statements about how nature behaves.

In most cases, laws are formulated mathematically, and they can be applied to support other discoveries, for example, in biochemistry or biophysics. Laws are generally discovered by observations; therefore, they stem from empirical evidence. Physics educators' task is to reflect on these methods while delivering these constructs to students as closely as possible. Developing empirical-mathematical models of the laws of physics emerges as the most profound method. While mathematical laws are considered to have absolute certainty, scientific laws can be questioned, contradicted, restricted, or modified by future observations and investigations (Fischbein, 1987).

Laws of physics, when expressed using algebraic symbols, can take various forms. The complexity of the structures depends on the mutual relationships of the quantities that constitute the law. A vast majority of the laws in physics are expressed in the forms of multivariable formulas. A science formula is defined as a concise way of expressing information symbolically or a general construct of a relationship between given quantities (Rautenberg, 2010). Formulas are context-related and thus are restricted to describe a specific phenomenon or its parts. However, they can be classified according to how their parameters' behavior is expressed mathematically. For example, if a direct proportionality is observed $f(x) = mx$ then structures of linear functions are applied. Formulas like $v(\omega) = R\omega$, $P(h) = \rho gh$, or $p(v) = mv$ can be interpreted scientifically following the meanings of the constant rates of change called also slopes when graphing is concerned. If an indirect proportionality is observed, then rational functions can be considered. For example, $R(A) = \frac{\rho l}{A}$ or $\alpha(I) = \frac{\tau_{Net}}{I}$. A more detailed classification of formulas concerning the complexity of the covariation is included in Chap. 5.

Laws expressed as algebraic functions can be regarded as dynamic algebraic structures that can reveal more details about phenomena. How students perceive formulas versus algebraic functions will be discussed in Part III of the volume and especially in Chaps. 12 and 13. Laws contain quantities that represent measurable entities. Quantities are called parameters when embraced in algebraic entanglements. Until the middle of the nineteenth century, algebra was almost identical to solving algebraic equations, particularly polynomials (Koch, 2000). The concepts of algebraic functions were not extensively applied in mathematics, and perhaps, they were not applied extensively in physics. For instance, Newton's law of universal gravity, constructed as a mathematical structure, was called a formula $F = \frac{Gm_1m_2}{d^2}$, and it has been used similarly nowadays to calculate one of the quantities labeled as either m_1, m_2, d, or F. Such application of this fundamental law of gravity is exercised widely and can be found in traditional physics textbooks. However, this formula can be perceived as an excellent example of a rational function of one variable if the parameter representing the distance between the masses' centers, d, is considered a variable and the remaining parameters are constant. Following this condition, the formula can be expressed as $F(d) = \frac{Gm_1m_2}{d^2}$, and its functional embodiment is illustrated in Fig. 1.1. While in mathematics, a graph position and appearance in XY coordinates are usually determined by the allowable function domain, position of $F(d)$ in F vs. d coordinates is determined by the restrictions that stem from its scientific image.

For example, to graph $F(d) = \frac{Gm_1m_2}{d^2}$, each mass can be considered 1 kg, and the distance d between the centers of the masses can be chosen as $d \geq 1$ m. Using this condition, the force of the mutual attraction is expressed in terms of the universal gravitational constant G; $F(d) = \frac{G(1\,\text{kg})(1\,\text{kg})}{d^2}$.

The graph illustrates that the force per unit of distance decreases, and the rate of the decrease can be computed by taking the derivative with respect to the distance: $F'(d) = \frac{-2G(1\,\text{kg})(1\,\text{kg})}{d^3}$. The graph also illustrates that the force decreases more

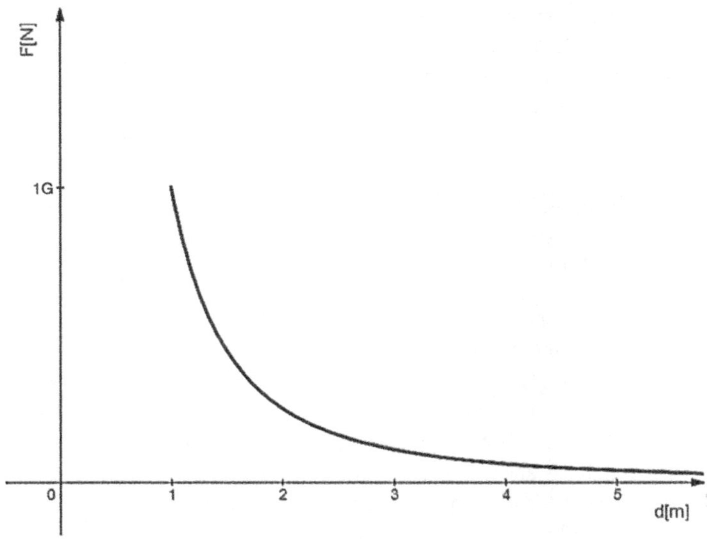

Fig. 1.1 Force of gravitational attraction between two masses being moved away; $d \geq 1$ m

rapidly for 1 m $\leq d \leq$ 2 m, and the rate of change and the force values are getting near zero when $d \to \infty$. For a considerable significant distance, the magnitude of the force can be approximated by employing limits that prove that the force is near the value of zero $F(d \to \infty) = \lim_{d \to \infty} \frac{Gm_1 m_2}{d^2} = 0$.

The law can also be investigated considering the distance between the objects as a constant parameter and the masses considered varying parameters. For instance, establishing the mass $m_1 = m_2 = m$, results in $F(d) = \frac{Gmm}{d^2} = \frac{Gm^2}{d^2}$. The function can be presented as $F(m) = \frac{G}{d^2}(m^2)$, and by considering e.g., $d = 1$ m, the general algebraic form can be simplified to $F(m) = Gm^2$. The formula is now a polynomial function of the second degree (Fig. 1.2).

The force is increasing due to an increase of masses, and the rate of increase of the force per unit of mass change is represented by a linear function $F'(m) = 2Gm$.

Both functions $F(d) = \frac{Gm_1 m_2}{d^2}$ and $F(m) = \frac{Gm^2}{d^2}$, when sketched in respective coordinates, evoke different inquiries about how the forces of mutual attraction change based on the functions' structural attributes. By considering different parameters as variables, different covariation is formulated, and its graphical representations take different forms, yet they are still about investigating the same phenomenon of gravitational attraction. By explicitly labeling the parameter that changes, its effects on the quantity of interest called the dependent parameter can be easily identified and its scientific behavior examined.

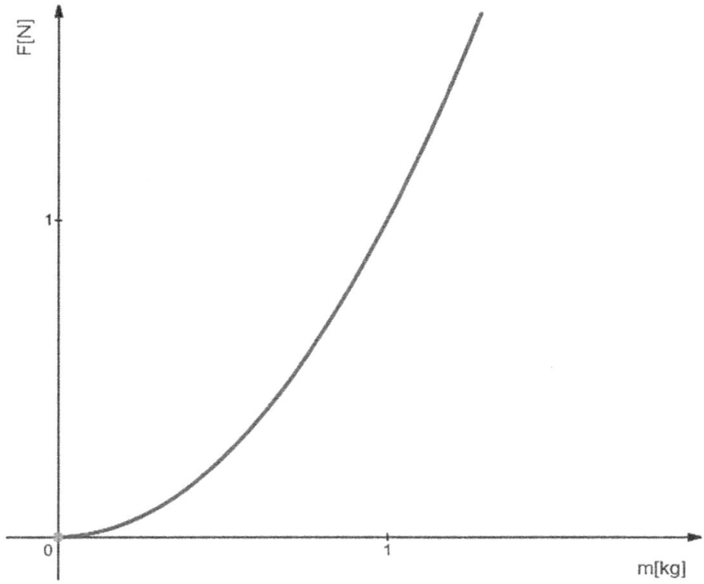

Fig. 1.2 Graph representing gravitational attraction between variable masses separated by 1 m

Another example of a physics law that can be converted to a one-variable function is the law of conservation of momentum (LCM) discovered and formulated by Isaak Newton. When expressed as equity of the initial and final momentum, the LCM takes the general form of $\sum_{n}^{i=1} \vec{p_i} = \sum_{n}^{i=1} \vec{p_f}$, and this form of the law dominates physics resources. A handful of problems can be found in literature where one of the unknown quantities (either mass, velocity, or momentum) is considered an unknown and all remaining parameters constant. While the earlier discussed form of the law of universal gravity resembled a function at first, the mathematical structure of the LCM does not resemble a function. Thus, the covariate relation between its parameters cannot be explicitly determined except that the left and the right sides must be numerically and vectorially identical. Analyzed by algebra rules, the law is merely a statement of equity of two sides of sums of algebraic expressions. In the absence of external forces, the total amount of kg·m/s remains constant before, during, and after the collision; thus, the graph of the system's total momentum versus time must represent a horizontal line (Fig. 1.3).

Can the law be converted to a one-variable covariate function? Such modification is possible. Once converted to one variable function, its behavior can be analyzed using either its graph or algebraic representation. For example, when considering that the two moving objects of masses m_1 and m_2 collide and got entangled after the collision, one parameter established as a variable can be used to theoretically determine the direction of motion of both masses after the impact. The direction of the

Fig. 1.3 Graph of the total momentum of two objects before, during, and after the collision

objects' motion when they are coupled can be derived from a general LCM and represented by:

$$\overrightarrow{v_o} = \frac{\overrightarrow{v_1}m_1 + \overrightarrow{v_2}m_2}{m_1 + m_2} \tag{1.1}$$

Suppose that one is interested in exploring the direction of the joint velocity of both objects after a perfectly inelastic collision when the mass m_1 can take various magnitudes and the velocities remain constant in each trial before the collision. To convert the problem to a one-variable relation, restrictions on its parameters must be made; thus the velocities of both objects remain constant; e.g., each is of a magnitude of 2 m/s as well as the mass m_2 must be constant, suppose $m_2 = 1$ kg. By these restrictions, the only variable parameter is the mass m_1. With that condition, the expression (1.1) takes the following covariate form

$$\overrightarrow{v_o}(m_1) = \frac{2m_1 - 2}{m_1 + 1}. \tag{1.2}$$

This rational function can be used to predict the direction of motion of the coupled objects after the collision and then to verify it using a simulated experimental setup (e.g., Fig. 1.4). The formulated covariate expression (Eq. 1.2) explicates the condition related to the total system mass when the coupled objects stop after the collision and when they move to the right or left after they are coupled. Students realize that to predict these behaviors, it is sufficient to analyze the numerator of the rational function because its denominator takes only positive values. While the simulated experiment allows only a definite range of magnitudes of the velocities and masses, the rational function form allows employing limiting case analysis and

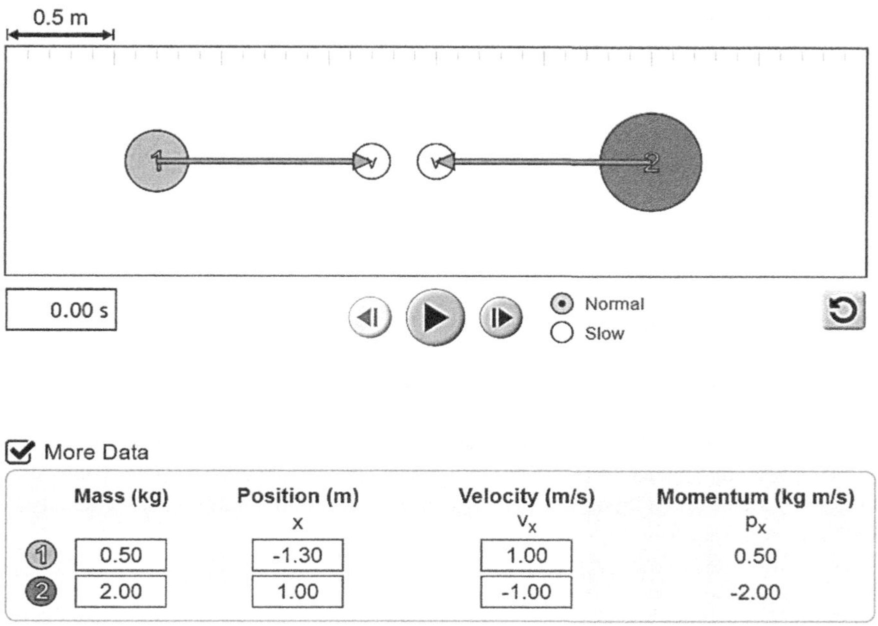

Fig. 1.4 Snapshot of an experiment investigating the law of conservation of momentum. (Source: Phet Interactive Simulations (n.d.))

hypothetically predict the systems' speed if the values of m_1 are considerably high. Such a task can be accomplished by taking the limit of the function (1.2) with respect to m_1 when $m_1 \to \infty$. Thus $\vec{v_o}\left(m_1 \to \infty\right) = \lim_{m_1 \to \infty} \frac{2m_1 - 2}{m_1 + 1} = \frac{2m}{s}$. The simulated experiment provides ways to verify the predictions using other parameters arrangements; the mathematical reasoning extends the analysis beyond the boundary of the lab setup.

When converted to covariate representations, a high diversity of algebraic forms of laws can shed light on different parts or phases of the phenomena and enrich its conceptual understanding. Upon reclassifying some quantities using the language of mathematics as independent, dependent, or constant according to their restricted behavior, such expressions are transferred to richer forms whose attributes can serve as a basis for employing deeper hypothetical thinking. These modifications add dynamics to the analysis and help zoom into how one quantity affects the other beyond the phenomenon classroom boundaries.

While most physics laws are expressed using the language of mathematics, some have been expressed qualitatively. For instance, Lentz's law that describes the direction of induced electric current in a solenoid due to a magnet's movement is an example of such a form that is often supported graphically. Another example can be Kepler's law for a planetary motion that states that the orbit of a planet is an ellipse with the Sun at one of the ellipse's focal points. This law is also supported by graphical representations using the image of an ellipse and its geometry. As some physics laws are formulated qualitatively and the others quantitatively, these representations often complement each other to amplify their meaning and support understanding.

Can abstract laws of mathematics be brought and applied to support physics understanding? Laws of mathematics are being applied for algorithmic operations at large. However, attempts are and can be made to include others. One of the laws commonly used in mathematics is the *transitive law* which says that if $a \times b = c \times b$, then one can infer that $a = c$. Attempting to apply this law in physics, one can state that if gravity acting on two different objects under the same intensity of the gravitational field is the same, thus if $m_1 \times g = m_2 \times g$, then the masses of the objects must be the same $m_1 = m_2$.

Examples of physics laws expressed qualitatively and quantitatively are found in any physics section. All laws reflect scientific rigor and a tremendous amount of work that their discoverers invested in bringing them to life. While the laws describe more significant spectra of reality, their subparts called *principles* can be considered building blocks of the laws. The principle as describing the nature will be discussed in the section that follows.

1.3 Principles and Their Relations to Laws

Principles are like laws, but they are usually less general than laws and refer to fundamental relationships between more than one physical quantity within a given phenomenon (Giancoli, 2008). Principles constitute a part of physics knowledge like axioms in mathematics. Principles are also defined as inherently clear statements used to guarantee the truth of the theorems from which they were derived (Easwaran, 2008). Since principles support a more comprehensive description of laws, they are often a part of the law, illustrated in Fig. 1.5.

Physics provides many examples of supportive roles of principles. For instance, the principle of wave superposition based on the algebraic property of function additivity supports a quantitative description of interference. If periodic functions are used to model the motion of the particles of the wave, then when two waves interact, the superposition principle informs that the resulting wave function is the sum of the two individual wave functions. The principle of wave superposition is often described graphically, and the resulting patterns are classified according to the

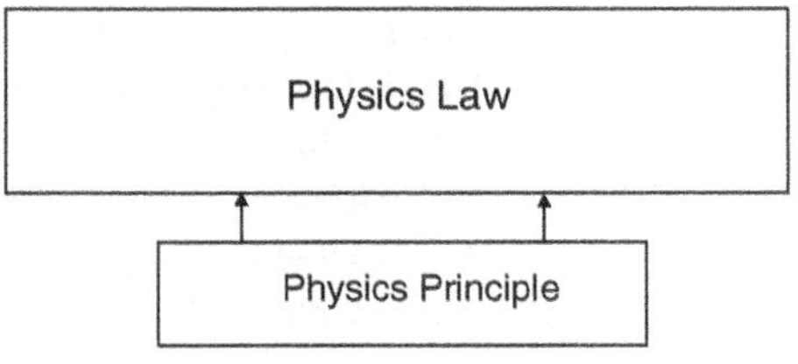

Fig. 1.5 Relation between principles and laws in physics

law of interference. While in high school practice, this principle is often introduced conceptually and embedded in graphical representations, it can also be supported by combining periodic functions expressed algebraically, presenting concurrently a gateway to more advanced Fourier analysis. The amplitude of interference is interpreted using a general law of interference, and consequently, the resulting patterns are classified as constructive, destructive, and totally destructive. The principle supports the quantification process, and the results are classified according to the law of interference. In its different yet congruent manner, the principle of superposing is also used in mechanics to compute the net force when two or more forces act on a particle simultaneously or when computing a net torque when various torques are exerted on an object simultaneously. The principle of superposition, in these contexts, enhances the process and advises that to find a net force or torque, a vectorial sum of all forces or torques acting on the object must be computed. Ultimately, using these principles can lead to formulating the operational form of Newton's laws of motion. It can be further inferred that the result of applying the principle of superposition can be called the action on the object and constitute one of the sides of Newton's law when it is expressed algebraically $\sum_{k}^{i=1} F_k = ma$. The right side expressed as a product of the object's mass, and acceleration can be called the result of the action. Interchanging these sides will still produce the same final answers, and students are usually given freedom in this regard. Literature does not set preference over the sides' order. However, from a cause-and-effect viewpoint, each representation $\sum_{k}^{i=1} F_k = ma$ and $ma = \sum_{k}^{i=1} F_k$ generates, to some extent, different scientific interpretations, even though the forms of the law are congruent.

Archimedes principle is yet another commonly taught physics principle. While it is used to quantify the buoyant force, it supports the law of floating, and it also aids the algebraic formulation of Newton's law to learn about an object's state of motion in the fluid. The Archimedes principle subsidized by the buoyant force's computations emerges as the building block to constructing the net force and, consequently, finding the object's acceleration. Thus, in describing an object's motion in fluids, two principles, Archimedes' principle and superposition of forces, can be considered supporting principles of Newton's second law when used to describe an object's behavior in that fluid. Principles are not parts of the laws' description; they support the law's formulations and enhance their applications, especially when quantification is concerned. Although not examined in physics research, making students aware of the contributions of these building blocks to the content of physics may have a profound effect on enhancing the pathways for their understanding.

The algebraic form of a law or principle is one side of the coin; interpreting the relation qualitatively and providing a supporting theory for its truthfulness is the other side that can evidence conceptual understanding. Theories are sometimes mistakenly considered theorems. The following section attempts to shed light on the differences between these two constructs and their relations to laws and principles.

1.4 Theories and Laws

The theory is described as an invented explanation of natural phenomena formulated using scientific methods and accepted fundamental principles or hypothetical deductive testing (Haig, 2018). The theory is also described as a qualitatively expressed testable prediction or hypothesis often accompanied by a great precision (Giancoli, 2008). A theory is not a random thought or an unproven concept. Theory needs to be experimentally verified to be accurate to speak the behavior of future similar events. To develop a scientific theory, a scientist must ask appropriate questions, design experiments, and draw appropriate conclusions from the results. While laws and principles describe the phenomena, theories are usually formulated verbally, and they explain the phenomena rather than describe them. To have a comprehensive understanding of a phenomenon, one needs to understand the law's formulation and its surrounding theory. Due to possessing an explanatory nature, theories are not formally proven; however, theory and data must be interwoven in a relationship, and the relationships must be rational and logical (Simon, 2012).

The relationships between theory and data must lead to correct conclusions about the extent to which those data support the theory. The physicist's task is to make explicit that the intuitively employed principles in justifying corresponding theories are correct (Giere, 2010). In this sight, taking data and analyzing their patterns can be used to verify the theory, but it is also applied to formulate a law quantitatively or qualitatively. Systematically taken data with clearly defined variable parameters can constitute the simplest form of an algebraic function, which can lead to enacting a formula called a law.

Laws and theories are both the products and tools of science, but each has a distinct heritage and role. One does not become the other (McComas, 2013). How law and theory can complement each other during class discussion illustrates the following example. Suppose that a system of objects possesses kinetic and potential energy and that the system satisfies the condition of having the total mechanical energy conserved. The law of conservation of mechanical energy states that at any time instant, the sum of the kinetic and potential energies of the system remains unchanged, which symbolically can be expressed as $K_{i\,sys} + U_{i\,sys} = K_{f\,sys} + U_{f\,sys}$, where K and U represent kinetic and potential energies, respectively. The law can be supported by the theory that energy cannot be created nor destroyed. Theory similarly to theorem can also be defined as a collection of axioms or definitions used to support a phenomenon's behavior. In general, an axiom is primarily used in mathematics, and it is a statement that is true to serve as a premise for reasoning (Duran-Guerrier, 2008).

Theory can elicit a model, primarily when theory is concerned with a simplified and idealized physical system or phenomenon (e.g., gas laws). A physical model usually mediates between scientific theory and reality. In contrast, laws are well established in physics education; theories that explain the laws perhaps not deliberately play a supportive role in physics teaching. Such relation of laws and theories might diminish the potential to develop students' deeper scientific reasoning and the phenomena understanding. When used in their mathematical forms, physics laws provide means for quantifications; they do not explain the phenomenon, nor can they justify students' conceptual understanding adequately. For instance, when asked to explain Newton's second law, students often claim that the law informs that force is equal to mass times acceleration. Such a statement does not explain conditions for accelerated motion, but it is merely a verbal description of the law. Explanation of the law demands an understanding of the actions and object's mass on its acceleration. The law is constructed as equity of action determined by the amount of net external force F acting on the object and its observable state of speeding up or slowing down, known as acceleration.

1.5 Theories and Theorems

There is a difference between theory and theorem; while theorems are not often used in physics, it is worthwhile to discuss these entities as well. Theorems are predominantly used in mathematics and defined as general statements not self-evident but proved by a chain of reasoning from a set of axioms established by mathematical logic (Loveland, 2016). A fundamental theorem of calculus (FTC) that states that change of function values is equal to accumulation under its derivative is one example of a mathematical theorem that is symbolically expressed as $\int_b^a f(x)dx = F(b) - F(a)$. This theorem, while not explicitly highlighted, is often applied in physics. For instance, in algebra-based physics courses, students calculate the area under the object's velocity-time graph and conclude its displacement. The fundamental theorem of calculus can also lead to deriving an integral function to compute, for example, a specific object's position when the initial position along with the velocity function is provided: $x(t) = x(0) + \int_t^0 v(t)dt$. It can also support finding the object's momentum at a specific time instant, t, when a variable impulse of force, $F(t)$ acts on it; $p(t) = p(0) + \int_t^0 F(t)dt$. Rich physics contexts can also be used to model the mathematical underpinnings of the FTC in calculus courses (Sokolowski, 2021).

Another example of a theorem is *the parallel-axis theorem* used to find the moment of inertia, I, of an object about an axis, parallel to the axis passing through the center of mass; $I = I_o + Mh^2$. The Mean Value Theorem is another example of a mathematical theorem that remains silent in physics that can enrich analyzing systems in physics. This theorem allows finding, for example, time instants when an object's instantaneous velocity is equal to its average velocity on the interval where the position function is differentiable.

Mathematical theorems support physics quantification and increase the set of tools to deploy to solve physics problems or enrich the inquiry. While theorems are statements often expressed symbolically, theories explain phenomena qualitatively.

The preliminary synthesis of the leading physics constructs illustrates that being derived empirically or hypothetically each contributes to physics knowledge in its own and distinctive right. While principles often support laws, theories initiate the formulation of the laws. It is beneficial to make students aware of each of these constructs' qualitative and quantitative roles to support holistic comprehension of physics discoveries. These constructs will arise while attempting to integrate mathematical reasoning into the context of physics that will be further deployed in this book.

References

Agnew, A. F., Bobe, A., Boskoff, W. G., & Suceavă, B. D. (2009). Gheorghe Ţiţeica and the origins of affine differential geometry. *Historia Mathematica, 36*(2), 161–170.

Bhushan, N., & Rosenfeld, S. (Eds.). (2000). *Of minds and molecules: New philosophical perspectives on chemistry*. Oxford University Press.

Duran-Guerrier, V. (2008). Truth versus validity in mathematical proof. *ZDM, 40*(3), 373–384.

Easwaran, K. (2008). The role of axioms in mathematics. *Erkenntnis, 68*(3), 381–391.

Feynman, R. P. (2005). *The pleasure of finding things out: The best short works of Richard P. Feynman*. Basic Books.

Fischbein, H. (1987). *Intuition in science and mathematics: An educational approach* (Vol. 5). Springer Science & Business Media.

Giancoli, D. C. (2008). *Physics for scientists and engineers with modern physics*. Pearson Education.

Giere, R. N. (2010). *Explaining science: A cognitive approach*. University of Chicago Press.

Haig, B. D. (2018) An abductive theory of scientific method. In *Method matters in psychology* (pp. 35–64). Springer.

Koch, H. (2000). *Number theory: Algebraic numbers and functions*. (No. 24). American Mathematical Society.

Koyré, A. (1957). *From the closed world to the infinite universe* (Vol. 1). Library of Alexandria.

Krupp, E. C. (2003). *Echoes of the ancient skies: The astronomy of lost civilizations*. Dover Publications.

Loveland, D. W. (2016). *Automated theorem proving: A logical basis*. Elsevier.

McComas, W. F. (Ed.). (2013). *The language of science education: An expanded glossary of key terms and concepts in science teaching and learning*. Springer Science & Business Media.

Oxford University Press. (2015). Conservation laws. In *Oxford dictionary of science* (7th ed., p. 212). Oxford University Press.

PhET Interactive Simulations. (n.d.). *The University of Colorado at Boulder*. Retrieved from http://phet.colorado.edu. September 2020.

Rautenberg, W. (2010). *A concise introduction to mathematical logic*. Springer.

Schwartz, M. (2012). *Principles of electrodynamics*. Courier Corporation.

Simon, H. A. (2012). *Models of discovery: And other topics in the methods of science* (Vol. 54). Springer Science & Business Media.

Sokolowski, A. (2021). Modelling the fundamental theorem of calculus using scientific inquiry. In F. K. S. Leung, G. A. Stillman, G. Kaiser, & K. L. Wong (Eds.), *Mathematical modelling education in east and west. International perspectives on the teaching and learning of mathematical modelling*. Springer. https://doi.org/10.1007/978-3-030-66996-6_36

Wigner, E. P. (1990). The unreasonable effectiveness of mathematics in the natural sciences. In *Mathematics and science* (pp. 291–306). World Scientific.

Young, H. D., & Freedman, R. A. (2004). *University physics with modern physics* (11th ed.). Addison Wesley.

Chapter 2
The Interface Between the Contents of Physics and Mathematics

2.1 Mathematics as a Language in Physics Classroom

The following sections shed light on how mathematics' structural domain can enrich physics understanding from research standpoints. While widely used, mathematics with its inaccessibility surpasses all the other scientific disciplines (Sfard, 1991). It is expected that students use mathematics to quantify scientific phenomena and apply mathematical reasoning to hypothetically predict phenomena behaviors. Quantifying the inputs/outputs of scientific experiments can be accomplished by applying simple algorithmic operations. However, activating students' hypothetical thinking to enrich the quality of inferences requires time and effort on the learners' and the instructors' sides.

According to research, scientists describe mathematics as a language of nature, which in practice is inadvertently reduced to routine calculations, especially when problems do not require a deeper mathematical insight. If the range of mathematics tools is not supported by its structural domain, it does not fully reflect what the phrase intends to mean practically. Such a view results in physics concepts delivered to students without their more profound and more thought-provoking scientific interpretations. Narrowing mathematics influx of using only its technical domain disables investigations from zooming into details of the scientific underpinnings and proving a new hypothesis or verifying laws. Thus, a different, more comprehensive view on introducing and applying physics concepts is sought. When defined, this task will require underlying these connections daily in the physics classroom while new concepts are introduced. Providing students with opportunities to explore such connections will position the students on the trajectory to endure the decisive role that mathematics plays in supporting their learning and, ultimately, on their desire to continue studying physics.

The effects of mathematics on discoveries in physics are unquestionable. As a source of models and abstractions, mathematics enables to obtain new insights into

A. Sokolowski, *Understanding Physics Using Mathematical Reasoning*,
https://doi.org/10.1007/978-3-030-80205-9_2

how natural phenomena behave and how nature operates (Justi & Gilbert, 2002). Thus, bringing it to the physics classroom appears as a task worthy of pursuing. Mathematics allows for symbolically expressing laws of nature and prompting redesigning already formulated laws in physics. For example, Dirac's formulated a wave function (Dirac's equation) that allowed him to discover a new particle, called positron (Farmelo, 2009). Rather than using purely scientific methods, including mathematics, with its powerful hypothetical–deductive tools, often empowered scientific discoveries and allowed for presenting them from different viewpoints. Enriching the teaching of physics concepts using the math structural domain can also initiate formulations of new representations of already known laws which can have profound effects on encouraging physics students to initiate their scientific inventions (e.g., Sokolowski, 2021). While appreciation of mathematics tools requires perhaps some level of maturity and experience, addressing the effects of mathematics on physics may also encourage students to delve deeper into the structures of mathematics.

2.2 Philosophy and the Substance of the Knowledge of Mathematics

The discussion of mathematical structures and their effects on physics reasoning depends on the complexity of these structures. However, it is worthy of discussing the nature and categories of mathematical knowledge as shaped by its development.

There exist two traditional assumptions concerning the nature of mathematics that (a) mathematical knowledge is a product of human knowledge and rationality that is secure and that (b) mathematical constructs, e.g., numbers, sets, and geometric objects, exist in some objective realm (Ernest, 1994). Consequently, there are two general paradigms of the philosophy of knowledge of mathematical (a) *epistemology* that is a study of beliefs about the acquisition of knowledge, and (b) *ontology* that is a study of beliefs about the nature of reality (Schraw & Olafson, 2008). Thom (1973, p. 204) claimed that "All mathematical pedagogy, even if scarcely coherent, rests on a philosophy of mathematics." Because the philosophy of mathematics has its classroom consequences (Steiner, 1987), the consequences must affect the content of mathematics and the teaching methods. The history of scientific discoveries shows that most physics laws and principles were discovered when mathematics was embraced in *epistemological* philosophy, which likely influenced its applications in science. This might perhaps explain why many physics laws are provided as static formulas that do not explicitly enhance exploratory and hypothetical didactical reasoning. Situating a framework for teaching mathematics following either *epistemology* or *ontology* affects many facets of mathematics pedagogy. These facets can include selecting the concepts for teaching, the objectives, teaching methodologies, didactical principles, learning theorems, models, and the theorems used in research (Steiner, 1987).

Since the middle of the 1980, mathematics philosophy took a different aim and moved away from complete epistemological systems toward ontology (Shapiro, 1997).

This shift influenced the content of mathematics curricula and the way mathematics has been taught. Consequently, it also polarized differently the way students use and perceive mathematics in science classes. As opposed to superficially doing mathematical operations, relational understanding of mathematics was a priority in which concept understanding and ideas came forth before the rules (Skemp, 1976). Such shift generated further needs for reclassifying the categories of mathematical knowledge taught to students and perhaps encouraged venues for exploring mathematical concepts in other disciplines such as physics.

2.3 Procedural and Conceptual Mathematical Knowledge

The shift in the philosophy of mathematics initiated the categorization of mathematical knowledge as *procedural* and *conceptual* (Hiebert & Wearne, 1986). These categories are the building blocks of competencies in mathematics education when developing the knowledge of concepts and procedures is the priority (Rittle-Johnson & Schneider, 2015).

Procedural knowledge encompasses familiarities with isolated facts or pieces of information, and it is applied to perform mathematical operations. Procedural knowledge does not explicitly develop students' conceptual thinking (Watson & Mason, 2006). An example of procedural knowledge can be familiarity with the symbols used in formulas and with tasks that lead to the symbols' desired configurations to solve problems. Mastering and retaining procedural knowledge is acquired by practice. For instance by solving equations of either algebraic, transcendental, or trigonometric forms.

Conceptual knowledge encompasses the understanding of structures of mathematics concepts (Hiebert & Wearne, 1986). It is gained by either (a) integrating elements of various knowledge within a discipline, (b) creating the relationships between prior and new knowledge, or (c) by linking knowledge of more than one discipline. Conceptual knowledge can also be thought of as an interconnected web of knowledge or a network in which linking relationships is as prominent as the discrete pieces of information (Hiebert & Lefevre, 1986). As learners progress with mathematics education, understanding conceptual math knowledge and abstract mathematical theorems and laws become the priority.

Procedural and conceptual math knowledge, see a diagram in Fig. 2.1, contribute to developing students' mathematical and scientific literacy and general STEM disposition. However, being able to possess and apply conceptual knowledge is inferior because it develops students' mathematical reasoning that reflects the needs of STEM education (Honey et al., 2014; Sokolowski, 2018) and the demands for twenty-first-century technological advances.

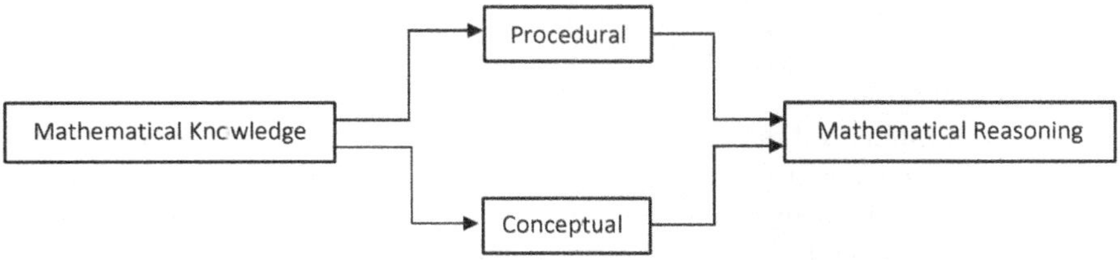

Fig. 2.1 Mathematical knowledge as categorized by mathematics education research

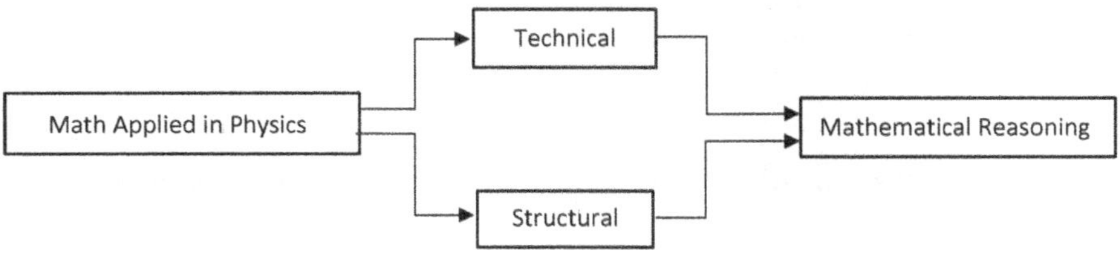

Fig. 2.2 Mathematical knowledge categorized by physics education research

The ability to reason mathematically is founded on managing both conceptual and procedural math knowledge with a superiority given to the conceptual counterpart.

2.4 Unifying Classification of Math Knowledge Used in Physics Education

While the terminologies used in math and physics to describe math knowledge do not change their interpretations, it is worthy of attempting to find parallelism. Physics research (Karam & Pietrocola, 2010) identifies two domains by which mathematics supports the meanings of physics concepts; (a) a technical domain that includes algorithmic operations and (b) a structural domain that includes these mathematical tools that help analyze system behavior in deeper theoretical analyses (Fig. 2.2).

While the technical domain requires employing students' technical skills, the structural domain requires activating students' algebraic reasoning and applying it to learn more about physical systems beyond evaluating formulas. Students' conceptual mathematics skills do not automatically convert into mathematics structural physics skills (Angell et al., 2008). This finding is perceived in physics education as one of the obstacles preventing students' physics knowledge from advancing to a higher level. It appeared beneficial to seek parallelism and unification of terminology used to also communicate such parallelism with the mathematics research community. Since the math knowledge used in physics courses correlates with that students learn in their math courses, mathematical knowledge used in physics must

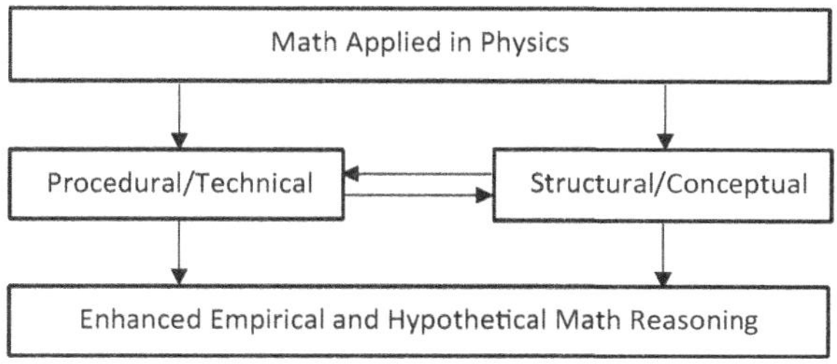

Fig. 2.3 Dual categorization of mathematics knowledge used in physics

have equivalences in mathematics education. As a result, a synthesis (see Fig. 2.3) has emerged.

Following the mapping, a general conclusion emerges that concepts and ideas encompassing *procedural math knowledge* in mathematics are considered *technical math knowledge* in physics. Concurrently, *conceptual knowledge of mathematics* can be categorized as *structural* in physics. Such mapping attempted to unify math knowledge categorization, and it is meant to assist math and physics educators in realizing the terminological similarities when addressing them to students.

In sum, efforts are made to enhance mathematical reasoning in physics by emphasizing *conceptual* – as defined in math, and *structural* – as defined in physics knowledge of mathematics. All four descriptors are being adapted in mathematics and physics education; therefore, to allow the transfer of terminology, all four descriptors will be used interchangeably yet with priority established by physics education communities.

2.5 Arrays of Applying Mathematics in Physics

The role of mathematics in physics has multiple dimensions. It (a) serves as a tool that constitutes pragmatic aspects, (b) acts as a language that represents communicative functions, (c) provides a logical framework for describing, ordering, and classifying physical processes and theories that encompass structural aspects (Krey, 2019). Delving deeper, mathematics as a tool provides technical/procedural algorithms, and mathematics as a language and framework provider supplies conceptual/structural aspects.

Considering the nature of physics students' difficulties to reason mathematically during physics inquiries, it seems that mathematics is underrepresented in communicative and structural aspects that stem from the domain of conceptual math knowledge. Possessing the ability to apply these structures can be acquired through familiarity with using various mathematical representations, mainly functions that can illustrate system dynamism, and interpreting these structures in the context of physics laws and principles. In this book, these aspects will be developed through identifying covariate relationships between selected phenomenon parameters and viewing that behavior by attending to the resulting covariate functions and their graphs. More specifically, such a structural domain of mathematics will be exercised by tools that help identify and interpret function attributes, such as function change, rates, limits, accumulation, or intercepts. Students learn and use functions in their mathematics courses as *domain-specific knowledge*. By being *domian-specific* that knowledge does not transfer to other subjects such as physics. Enabling this transfer of knowledge from one discipline within another requires reconstructing that knowledge in the new contexts and settings (Pugh et al., 2015). Research shows that such undertaking is complex, and it proves to be more challenging than many physics teachers anticipate (Orton & Roper, 2000). Following the research suggestions, such reconstruction requires teacher guidance in pinpointing critical phases or integration, especially in extracting scientific knowledge from the behavior and outputs of algebraic structures.

The research documented that procedural knowledge encompasses congruent structures and procedures in mathematics and physics. However, conceptual mathematics knowledge encompassing the structural domain does not have its equivalent counterpart in physics at hand. This means that to support the transfer of conceptual math knowledge to physics, math concepts must be reconstructed to be readily available to physics students. Such reconstruction is not about reteaching these concepts or letting students review these concepts on their own. To succeed, such reconstructions must provide a web of impulses that the students will be attracted to retain and apply to support their theories during analyses of natural phenomena. In contemporary research on physics education, such attempts are rare. While a wealth of theoretical theses about a need for such didactical enterprises can be found, a lack of projects is seen as the main reason for fragile progress. To effectively stipulate students' mathematical reasoning in physics, conceptual knowledge of mathematics and the structures that these concepts are built upon are vital in the physics classroom and central in formulating conceptual and algebraic models. Organizing physics learning experiences that gravitate toward such an aim is to transcend students' reasoning skills to a level that will enable them to make mental connections between phenomena behaviors and their corresponding mathematical embodiments. One of the fundamental methods the scientific community uses to uprise new ideas, theories, and laws is evaluating whether and how they cohere with existing knowledge (Thagard, 2007). Therefore, mathematical reasoning that can serve as a determinant to support predictions, hypotheses, and reconstructing or deriving physics knowledge will also need to adhere to existing norms. Some scientists (e.g., Branchetti et al., 2019) are more explicit with such description claiming that mathematics is much more than a language for dealing with the physical world; it is a source of models and abstractions which enable to obtain new insights into how nature operates. Since this book's goal was to propose didactical methods to enrich students' scientific inquiry by mathematical reasoning, an attempt to review mathematics concepts that can engage physics students in deeper mathematical reasoning will also be made.

2.6 Search for Tools and Methods

Mathematical reasoning or thinking has multiple definitions. Harel and Tall (1991) defined it as a process that enables learners to enrich and expand their ideas. Several mental processes such as conjecturing, generalizing, convincing, comparing, reversing, generalizing, explaining, justifying, and verifying can be induced (Mason, 1989). Mathematical reasoning can also include auxiliary processes such as calculating, solving, drawing and measuring. All these mental and physical activities can involve either procedural or conceptual knowledge or their blend, and as a result, new knowledge can be inferred. While there is specific variability in how these two constructs are defined, there is consensus that the relations between conceptual and procedural knowledge are often bi-directional and iterative. Thus, applying these attributes, conceptual knowledge can be derived from procedural, and procedures can be altered due to what conceptual knowledge is sought. Such bidirectional relation has its place in physics; algebraic computations can lead to unique inferences about the system behavior; formulating a theory can trigger specific algebraic procedures that will lead to a law formulation. Iteration also has its roots in scientific methods, which are entrenched in the procedures to reach a decision or a desired result by repeating rounds of analysis or a cycle of operations. A consensus has been reached that the knowledge of both categories, procedural and conceptual, affects mathematical reasoning development. The literature suggested two critical practices to enhance mathematical reasoning: (a) justifying and generalizing and (b) symbolizing, representing, and communicating. For some researchers (e.g., Kilpatrick et al., 2001), justifying as to "provide sufficient reason for" (p. 130) is a critical element of being fluent to reason mathematically. These mental processes can be exercised from analyses of covariational representations and many of such metal actions like generalizing, symbolizing, representing, and communicating, are included in STEM modeling frameworks designed for enhancing scientific inquiry in mathematics (Marginson et al., 2013; Sokolowski, 2018). The prospect of using algebraic representations in physics courses derived from natural phenomena can enrich students' physics and math skills. In an earlier study, Sokolowski (2015) found out that modeling activities that use scientific contexts in mathematics courses present math ideas in a way that makes the students view these ideas in a different more pragmatic perspective. Many mathematical representations: symbols, numbers, tables, diagrams, graphs, and algebraic expressions are being used to represent physical constructs. Beyond this representational role, mathematics allows for conceptual insight that might produce richer inferences (Quale, 2011). Therefore, to develop scientific literacy, physics should be taught to rely not only on observing experiments, but these experiments should provide bases for empirical-mathematical reasoning concluded with symbolically expressed behavior of selected experiment's parameters.

2.7 Mathematical and Scientific Reasoning; Are These Mental Actions Equivalent?

Enhancing physics inquiry by mathematical reasoning can take diverse paths depending on several factors such as the layer of embedded reasoning, the grade level, content taught, and the alignment of mathematics and physics curricula. Though, there are certain general factors and strategies that can help flesh out these paths.

This paragraph is to delve deeper into identifying conceptual and procedural knowledge and their effects on developing students' mathematical reasoning skills in physics. In concluding, a general link between mathematics knowledge learners acquire in mathematics and that needed in physics will emerge. These connections will be established through a preliminary analysis of the possible reasoning approaches developed and used in mathematics and physics.

Theorems, laws, and axioms in mathematics differ from these studied in sciences. They are introduced as context-free because they need to be universal to any subject application. Because of the lack of context, mathematics laws are presented very abstractly. Physics constitutes a natural framework for testing, applying, and elaborating mathematical theorems, methods, and concepts, or even motivating, stimulating, instigating, and creating all kinds of mathematical innovations (Tzanakis & Thomaidis, 2000). However, these methods are not widely exercised in mathematics; thus, students entering a physics classroom typically possess the tools, but they do not possess the skills how to use these tools to support scientific analyses. To initiate the transfer of knowledge, mathematical symbols, concepts, and laws that students learn in their math classes should be refreshed and augmented in physics considering their purpose of being used in physics. For this to sustain, the semantics that informs how the mathematical symbols are interpreted in math classes should also be transferrable to physics classrooms and interpreted congruently with their mathematical meanings. The final products of such actions should lean on scientific interpretations.

2.8 Synthesis of Students' Challenges with Math Knowledge Transfer

Research shows that students can easily activate their procedural mathematics knowledge in physics classes, but they face difficulties activating more sophisticated conceptual mathematical knowledge (Wilcox et al., 2013). This can be explained in the way how mathematics knowledge is delivered to students. Procedural math skills such as calculating, solving, drawing, or measuring, can be easily transferred to physics due to the nature of these activities classified as domain-specific developed and used in a similar nature in mathematics as in physics. Learners do not encounter difficulties in transferring procedural *domain-specific* math knowledge because they construct the understanding of these concepts within similar disciplinary norms in both subjects. Calculating or solving for a variable are elements of procedural math knowledge and such knowledge is transferred effortlessly from mathematics to other subjects and also to physics. The only element of the new knowledge encountered in physics is the syntactic (not conceptual) understanding of physics symbols if calculating or solving for a parameter is involved. While working on such physics problems, intuitive mathematics knowledge, symbolic forms, and interpretive devices are not put into play (Tuminaro & Redish, 2007). Transferring conceptual math knowledge is different. The barriers that arise when transferring conceptual math knowledge to physics classes are researched, and these difficulties are well documented by studies encompassing multiple countries, curricula, and classroom settings (Pospiech, 2019). A handful of these studies report these difficulties as shortfalls of students' abilities to activate conceptual math knowledge to support physics reasoning (e.g., Fraser et al., 2014). Planinic et al. (2013) proposed three categories of possible sources of these difficulties (a) either that the required resource does not exist (students did not learn required math concepts in their math classes), or (b) the resource exists but is not activated due to the incorrect framing of the problem (this could result from a misalignment of the terminology used) or (c) the resource is activated, but its mapping to the problem is not appropriate (this could result from a misalignment of symbols or structures). Research provides recommendations on eliminating these barriers; Redish (2005) suggested augmenting mathematics tools so that their syntaxes fit into these used in physics. Others found out that transfer is more likely to occur when students realize a given idea in at least two diverse contexts or receive metacognitive scaffolding (see Hammer et al., 2005). Branchetti et al. (2019) used a model developed by Uhden et al. (2012) to design a teaching tutorial to help mathematics and physics university students with knowledge transfer. Wilcox et al. (2013) developed an analytical framework to assist instructors and researchers in aligning students' difficulties with specific mathematical tools when solving problems of the upper physics division. Kuo et al. (2020) found out that students sought coherence between formal mathematics and conceptual understanding and suggested that mathematical sense making – the practice of seeking coherence between formal mathematics and conceptual understanding – is a critical element of physics problem-solving. diSessa (2008) suggested that (a) inquiry is a more successful way to develop ontological meanings than other methods and (b) that students can become much wiser with this kind of process and enhance future learning on that basis.

This synthesis illustrates that students' difficulties can be accounted for by a lack of parallel activities in physics courses that would activate their conceptual mathematical knowledge departing from the way they learned these concepts in math classes. It can be inferred that expecting that students activate that knowledge automatically without the teacher's guidance can be deceptive. Organizing physics learning experiences that gravitate toward knowledge transfer is seen as equipping students with reasoning skills to help them make mental connections between phenomena behaviors and interpretations of corresponding mathematical embodiments.

In sum, it is reasonable to say that obstacles with transferring math knowledge stem from a lack of parallelism between methods, approaches, and perhaps syntaxes that students learned in math and what the expectations in physics are. Establishing a conceptual framework that will attempt to elevate the challenges and allow this transition will be discussed and proposed in Part II of the book.

References

Angell, C., Kind, P. M., Henriksen, E. K., & Guttersrud, Ø. (2008). An empirical-mathematical modeling approach to upper secondary physics. *Physics Education, 43*(3), 256.

Branchetti, L., Cattabriga, A., & Levrini, O. (2019). Interplay between mathematics and physics to catch the nature of a scientific breakthrough: The case of the blackbody. *Physical Review Physics Education Research, 15*(2), 020130.

diSessa, A. A. (2008). A "theory bite" on the meaning of scientific inquiry: A companion to Kuhn and Pease. *Cognition and Instruction, 26*(4), 560–566.

Ernest, P. (1994). The philosophy of mathematics and the didactics of mathematics. In *Didactics of mathematics as a scientific discipline* (pp. 335–350). Kluwer Academic Publishers.

Farmelo, G. (2009). *The strangest man: The hidden life of Paul Dirac, quantum genius*. Faber & Faber.

Fraser, J. M., Timan, A. L., Miller, K., Dowd, J. E., Tucker, L., & Mazur, E. (2014). Teaching and physics education research: Bridging the gap. *Reports on Progress in Physics, 77*(3), 032401.

Hammer, D., Elby, A., Scherr, R. E., & Redish, E. F. (2005). Resources, framing, and transfer. In *Transfer of learning from a modern multidisciplinary perspective* (pp. 89–119). IAP.

Harel, G., & Tall, D. (1991). The general, the abstract, and the generic in advanced mathematics. *For the Learning of Mathematics, 11*(1), 38–42.

Hiebert, J., & Lefevre, P. (1986). Conceptual and procedural knowledge in mathematics: An introductory analysis In J. Hiebert (Ed.), *Conceptual and procedural knowledge: The case of mathematics* (pp. 1–27). Erlbaum.

Hiebert, J., & Wearne, D. (1986). Procedures over concepts: The acquisition of decimal number knowledge. In J. Hiebert (Ed.), *Conceptual and procedural knowledge: The case of mathematics*. Erlbaum.

Honey, M., Pearson, G., & Schweingruber, H. A. (Eds.). (2014). *STEM integration in K-12 education: Status, prospects, and an agenda for research* (Vol. 500). National Academies Press.

Justi, R. S., & Gilbert, J. K. (2002). Modelling, teachers' views on the nature of modelling, and implications for the education of modellers. *International Journal of Science Education, 24*(4), 369–387.

Karam, R., & Pietrocola, M. (2010). Recognizing the structural role of mathematics in physical thought. In *Contemporary science education research: International perspectives* (pp. 65–76). Pegem Akademi.

Kilpatrick, J., Swafford, J., & Findell, B. (2001). *Adding it up: Helping children learn mathematics* (Vol. 2101). National Research Council (Ed.)). National Academy Press.

Krey, O. (2019). What is learned about the roles of mathematics in physics while learning physics concepts? A mathematics sensitive look at physics teaching and learning. In *Mathematics in physics education* (pp. 103–123). Springer.

Kuo, E., Hull, M. M., Elby, A., & Gupta, A. (2020). Assessing mathematical sensemaking in physics through calculation-concept crossover. *Physical Review Physics Education Research, 16*(2), 020109.

Marginson, S., Tytler, R., Freeman, B., & Roberts, K. (2013). *STEM: Country comparisons: International comparisons of science, technology, engineering, and mathematics (STEM) education*. Final report.

Mason, J. (1989). Mathematical abstraction as the result of a delicate shift of attention. *For the Learning of Mathematics, 9*(2), 2–8.
Orton, T., & Roper, T. (2000). Science and mathematics: A relationship in need of counselling? *Studies in Science Education, 35*, 123–154.
Planinic, M., Ivanjek, L., Susac, A., & Milin-Sipus, Z. (2013). Comparison of university students' understanding of graphs in different contexts. *Physical Review Special Topics – Physics Education Research, 9*(2), 020103.
Pospiech, G. (2019). Framework of mathematization in physics from a teaching perspective. In *Mathematics in physics education* (pp. 1–33). Springer, Cham.
Pugh, K. J., Linnenbrink-Garcia, L. I. S. A., Phillips, M. M., & Perez, T. O. N. Y. (2015). *Supporting the development of transformative experience and interest* (pp. 369–383). AERA.
Quale, A. (2011). On the role of mathematics in physics. *Science & Education, 20*(3–4), 359–372.
Redish, E. F. (2005). Changing student ways of knowing: What should our students learn in a physics class. *Proceedings of World View on Physics Education*, 1–13.
Rittle-Johnson, B., & Schneider, M. (2015). Developing conceptual and procedural knowledge of mathematics. In *Oxford handbook of numerical cognition* (pp. 1118–1134). Oxford University Press.
Schraw, G. J., & Olafson, L. J. (2008). Assessing teachers' epistemological and ontological worldviews. In *Knowing, knowledge, and beliefs* (pp. 25–44). Springer.
Sfard, A. (1991). On the dual nature of mathematical conceptions: Reflections on processes and objects as different sides of the same coin. *Educational Studies in Mathematics, 22*(1), 1–36.
Shapiro, S. (1997). *Philosophy of mathematics: Structure and ontology*. Oxford University Press on Demand.
Skemp, R. R. (1976). Relational understanding and instrumental understanding. *Mathematics Teaching, 77*(1), 20–26.
Sokolowski, A. (2015). The effects of mathematical modelling on students' achievement-meta-analysis of research. *IAFOR Journal of Education, 3*(1), 93–114.
Sokolowski, A. (2018). Formulating conceptual framework for multidisciplinary STEM modeling. In *Scientific inquiry in mathematics – Theory and practice* (pp. 53–62). Springer.
Sokolowski, A. (2021). Enabling covariational reasoning in Einstein's formula for photoelectric effect. *Physics Education, 56*(3), 035029.
Steiner, H. G. (1987). Philosophical and epistemological aspects of mathematics and their interaction with theory and practice in mathematics education. *For the Learning of Mathematics, 7*(1), 7–13.
Thagard, P. (2007). Coherence, truth, and the development of scientific knowledge. *Philosophy of Science, 74*(1), 28–47.
Thom, R. (1973). Modern mathematics: Does it exist? In A. G. Howson (Ed.), *Developments in mathematical education* (pp. 194–209). Cambridge University Press.
Tuminaro, J., & Redish, E. F. (2007). Elements of a cognitive model of physics problem solving: Epistemic games. *Physical Review Special Topics – Physics Education Research, 3*(2), 020101.
Tzanakis, C., & Thomaidis, Y. (2000). Integrating the close historical development of mathematics and physics in mathematics education: Some methodological and epistemological remarks. *For the Learning of Mathematics, 20*(1), 44–55. Retrieved January 18, 2021, from https://eric.ed.gov/?id=ej607175
Uhden, O., Karam, R., Pietrocola, M., & Pospiech, G. (2012). Modelling mathematical reasoning in physics education. *Science & Education, 21*(4), 485–506.
Watson, A., & Mason, J. (2006). Seeing an exercise as a single mathematical object: Using variation to structure sense-making. *Mathematical Thinking and Learning, 8*(2), 91–111.
Wilcox, B. R., Caballero, M. D., Rehn, D. A., & Pollock, S. J. (2013). Analytic framework for students' use of mathematics in upper-division physics. *Physical Review Special Topics – Physics Education Research, 9*(2), 020119.

Part II
Designing Learning Environments to Promote Math Reasoning in Physics

Chapter 3
Modeling as an Environment Nurturing Knowledge Transfer

3.1 Scientific Modeling and Models

The case studies proposed in this book (Part III) conclude with models, often expressed as covariate algebraic expressions with modeling considered a process that will lead students to formulate these algebraic models. This section delves deeper into the learning advantages that stem from immersing students in modeling activities to pursue math knowledge transfer. A model in science has many definitions. It is perceived as a surrogate object or a conceptual representation of a part of reality that enables its understanding and theory formulation. As theory explains a phenomenon behaviour, the model aids in understanding specific aspects of the phenomenon. The primary purpose of formulating models is enabling an epistemological link between realities, theory, and law (see Fig. 3.1).

A model is formulated by applying a suitable strategy, and it is to reflect on the relations of system parameters in the search for a deeper understanding. Model is also called a kind of analogy or mental image of the phenomena expressed in other structures that we are already familiar with (Giancoli, 2014). To serve its purpose, the model needs to be simple and provide a structural similarity to the phenomena under investigation. While the theory is based on certain assumptions, the model is typically free from assumptions and is restrained in its representation.

Modeling is considered the most popular and effective way of developing algebraic representations to interpret system behavior (Selden & Selden, 1992). While modeling is the action, the model is the product of that action. Modeling activities bring students closer to science's epistemic view and warrant more insightful ideas about the scientific inquiry that they can further deploy when working on other scientific investigations or solving real problems. Physics is a modeling enterprise, and as such, it should train students to become competent modelers and interpreters of these models (Angell et al., 2008).

A. Sokolowski, *Understanding Physics Using Mathematical Reasoning*,
https://doi.org/10.1007/978-3-030-80205-9_3

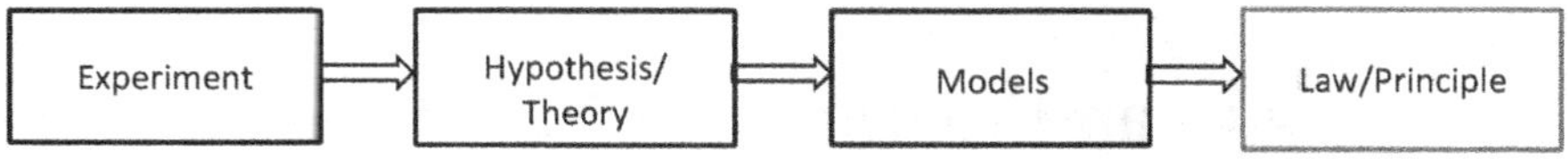

Fig. 3.1 Models as mediators between experiment, theory, and law

Modeling allows blending representations naturally, and it is accepted as an essential tool in science and mathematics education at all levels (Lesh et al., 2010). There are more descriptions of modeling found in research. Odenbaugh (2005) defines modeling as a collection of cognitive strategies formulated to pursue scientific inquiry aims. Wilkinson (2011) perceived modeling as an attempt to develop an understanding of elements of a system, considered a part of reality, their states, and interactions with other elements.

Models of natural phenomena similar to representations can take different forms; conceptual, mental, verbal, physical, statistical, logical, graphical, etc. Research identifies models created using mathematical structures as the most prominent and most frequently deployed to modeling physics activities (Redish & Kuo, 2015). While being formulated, physics constructs mediate between more than one type of model before reaching their final representational forms. Creating mathematical models that allow both predictions and theory is perceived as a primary physics education goal (Hestenes, 1987). When mathematical representations are the final products, they must accord with formulating empirical-mathematical models suggested by research in mathematical modeling. Such alignment will nurture math knowledge transfer.

One of the essential advantages of using models is that they reduce a need to transfer domain-specific knowledge by providing meaningful ways for students to construct, explain, describe, manipulate, or predict patterns and regularities associated with the situation under investigation (Michelsen, 2005). The development of the competency of using mathematics reasoning in science is not just about applying mathematics algorithms. The competency is developed by merging interpretations of the phenomena behavior with attributes of algebraic functions. Physics as an experimental science provides multiple opportunities for organizing such learning venues.

3.2 Modeling Cycles in Physics Education

Traditional development of the competency of using mathematics in science is about applying mathematics algorithms. The book is posited to change this perception. Using mathematics can be enriched by merging interpretations of the phenomena behavior with mathematical representations where modeling is a mediator during these enterprises. While the modeling products can take diverse forms, modeling that concludes with the formulation of an algebraic function representing a formula will be the priority in this book. Being such, modeling applied in experiments is to generate the basis for developing algebraic representations of the part of reality. Applying scientific methods that delve deeper in formulating and testing hypotheses and merge math and physics reasoning appears a rich learning experience. Such tasks enhance the students' scientific inquiry skills as they have the prospective to make the concepts of science and mathematics more tangible and more attractive to explore and understand.

There exist several modeling cycles in the physics education literature that assist in organizing modeling activities. Hestenes (1995) proposed to initiate the modeling process by isolating a part of a situation called the phenomenon. More detailed phases of these enterprises, such as purpose and validity, were included in that process, and all supported the model formulation. Analysis and justification of the enacted model concluded the modeling cycle. Students' depth of reasoning that emerges from this modeling depends on the specific nature of the investigations at hand. While it was assumed that models, as the products of such investigations, could take various forms, algebraic representations were assumed to be one of them.

Redish and Gupta (2009) developed a modeling cycle that explicitly linked algebraic representations with a physical system by mental actions such as processing, interpreting, and evaluating. Modeling appeared as the catalyst of transferring properties of the physical system into a mathematical representation. Such formulated representations were to be interpreted and further appraised for their accuracy to reflect on the physical system's behavior under investigation.

More recently, Uhden et al. (2012) developed a scheme how mathematics is used in physics and how it should enhance mathematical reasoning. The model was intended to provide a guiding framework when aspects related to mathematical reasoning in physics were concerned. It exemplified various levels of knowledge of mathematics to constructing mathematical-physical representations. The sophistication of using the tools of mathematics depended on the degree of mathematization endured with technical mathematical operations regarded as pure mathematics tools and structural skills regarded as students' capacity of employing mathematical knowledge for structuring physical situations. The interpretation process included a mathematical-physical model and was to enhance the scientific part of the analysis.

The modeling cycles discussed herein were to support instructors in designing lessons and laboratory activities. As such, they serve the general purpose of establishing a didactical framework. By being broadly defined, they can be applied to any grade level/physics section and help set up the modeling processes. According to research, there is a need for more detailed designs of lab activities that will illustrate classroom interactions while exemplifying mathematical reasoning. Thus, a general framework and specific content-related actions that will support these frameworks are compulsory. Reflecting on these actions and identifying the prompts that will activate students' math knowledge appeared a priority that will be discussed in the following sections.

3.3 Merging Mathematics and Physics Representations

Prain and Waldrip (2006) called for establishing a prominent role of representations in physics education to enhance scientific inquiry because they fit into students' learning styles and engage students to learn science. To reflect on this recommendation, in mathematics courses students are presented, for instance with various function forms followed by the *rule of five*, which suggests using five function representations: verbal description, symbolic representation, graphs, table of values, and mapping (Yerushalmy, 1997). Students are to use these representations and be able to transfer between them. Representations that resonate with students' realities are easily converted into mental impulses and stored in students' long-term memory. Being able to produce new representations enables deriving new theories. Prominent scientists made their discoveries by carefully selecting and analyzing existing representations and then inventing new (Cheng, 1999). Such prospect of tasks seems reasonable to apply in physics courses. More sophisticated enterprises can include applying the concept of limits, called also *limiting case analyses* that will be used more extensively in Chaps. 6 and 9. These analyses will enhance scientific inquiries in situations where direct evaluations of derived covariation relationships will not be plausible.

Mathematics plays an essential role in quantifying science and embracing it in concise mathematical language. Thus, being an expert in understanding and formulating scientific laws and theories and applying these constructs to solve physics problems requires recognizing essential elements between various scientific representations and their mathematical counterparts. Upon understanding these entanglements, learners are set to become versed in applying these entities to construct new laws or reconstruct laws that had already been discovered. Merging mathematical reasoning with its scientific counterpart means assuring coherence between interpretations and enabling transitions not only between various mathematics representations but also between their physics equivalent counterparts. Acting upon enabling such transitioning appears to strengthen understanding of empirical evidence.

Using the mathematical way of reasoning to enhance scientific inquiry is also the key to success in engineering programs where expert problem-solvers need to exploit opportunities to use blended conceptual reasoning (Fauconnier & Turner, 2003). Working on lab activities or solving physics problems often concludes with formulating algebraic models of the phenomena behavior. All these mental processes inherent in identifying prompts and reconstructing them in various representations merge into one coherent thought process. Diverse physics curricula comprising laws, principles, and theorems expressed qualitatively or quantitatively provide a wealth of opportunities for enacting these lines of inquiry. Research showed that reasoning, and especially this supported by empirical evidence, has proven difficult for physics students (Hammer, 1996); therefore, developing learning environments that will produce a web of stimuli that students can use to engage in these processes to reach fluency in this domain is needed. In Part II, Chap. 4, such a modeling scheme is proposed. The scheme provides a general design for lab activities, and lesson conducts that prioritize the phase of merging the methods of physics and mathematics.

References

Angell, C., Kind, P. M., Henriksen, E. K., & Guttersrud, Ø. (2008). An empirical-mathematical modeling approach to upper secondary physics. *Physics Education, 43*(3), 256.

Cheng, P. C. H. (1999). Unlocking conceptual learning in mathematics and science with effective representational systems. *Computers & Education, 33*(2), 109–130.

Fauconnier, G., & Turner, M. (2003). *The way we think: Conceptual blending and the mind's hidden complexities*. Basic Books. ISBN: 9780465087860.

Giancoli, E. (2014). *Physics: Principles with applications* (AP ed.). Pearson.

Hammer, D. (1996). More than misconceptions: Multiple perspectives on student knowledge and reasoning, and an appropriate role for education research. *American Journal of Physics, 64*(10), 1316–1325.

Hestenes, D. (1987). Toward a modeling theory of physics instruction. *American Journal of Physics, 55*(5), 440–454.

Hestenes, D. (1995). *Modeling software for learning and doing physics*. In Thinking physics for teaching (pp. 25–65). New York: Springer US.

Lesh, R., Galbraith, P. L., Haines, C. R., & Hurford, A. (2010). *Modeling students' mathematical modeling competencies*. Boston, MA: Springer US. doi, 10, 978-1.

Michelsen, C. (2015). Mathematical modeling is also physics—interdisciplinary teaching between mathematics and physics in Danish upper secondary education. Physics *Education, 50*(4), 489.

Odenbaugh, J. (2005). Idealized, inaccurate but successful: A pragmatic approach to evaluating models in theoretical ecology. *Biology and Philosophy, 20*(2–3), 231–255.

Prain, V., & Waldrip, B. (2006). An exploratory study of teachers' and students' use of multi-modal representations of concepts in primary science. *International Journal of Science Education, 28*(15), 1843–1866.

Redish, E. F., & Gupta, A. (2009). Making meaning with math in physics: A semantic analysis. *GIREP-EPEC & PHEC 2009*, 244.

Redish, E. F., & Kuo, E. (2015). Language of physics, the language of math: Disciplinary culture and dynamic epistemology. *Science & Education, 24*(5), 561–590.

Selden, A., & Selden, J. (1992). Research perspectives on conceptions of function: Summary and overview. In *The concept of function: Aspects of epistemology and pedagogy* (Vol. 25, pp. 1–16). Mathematical Association of America.

Uhden, O., Karam, R., Pietrocola, M., & Pospiech, G. (2012). Modelling mathematical reasoning in physics education. *Science & Education, 21*(4), 485–506.

Wilkinson, D. J. (2011). *Stochastic modelling for systems biology*. Boca Raton, FL: CRC Press.

Yerushalmy, M. (1997). Designing representations: Reasoning about functions of two variables. *Journal for Research in Mathematics Education, 28*(4), 431–466.

Chapter 4
Proposed Empirical-Mathematical Learning Model

4.1 Didactical Underpinnings of the Design

While the general design of the scheme aims at merging the math structural domain with physics concepts, other objectives such as sparkling students' interest in studying physics and encouraging their investigations were also considered. The empirical-mathematical modeling scheme was inspired by an earlier designed modeling cycle where physics aided the link between natural phenomena and their engineering and technology applications (Sokolowski, 2018). As the primary inquiry process remains scientific in both designs, the one presented in Fig. 4.1 highlights physics's understanding as the primary learning objective. Instructional suggestions on how to use it in practice follow.

The process is designed for organizing lab activities, however it can also serve as a backbone to developing instructional units that intend to immerse students in mathematical reasoning while exploring the physics concepts. The scheme consists of mental or physical actions, represented by the cells in the middle column of the diagram. These actions will reflect on the specific lab objective and goals. The cells on the right and left sides of the stem actions indicate phases when integrating mathematics and physics takes priority. When completed, the formulated mathematical model can be deployed to problem solving or it can constitute new knowledge derived from the investigations.

4.2 Description of the Learning Phases

The labelings in the first two cells as real or simulated contexts pertain to preparation and formulating the lab purpose. These tasks are to be prepared by the instructor. The remaining tasks in the stem actions reflect on the methods of scientific

A. Sokolowski, *Understanding Physics Using Mathematical Reasoning*,
https://doi.org/10.1007/978-3-030-80205-9_4

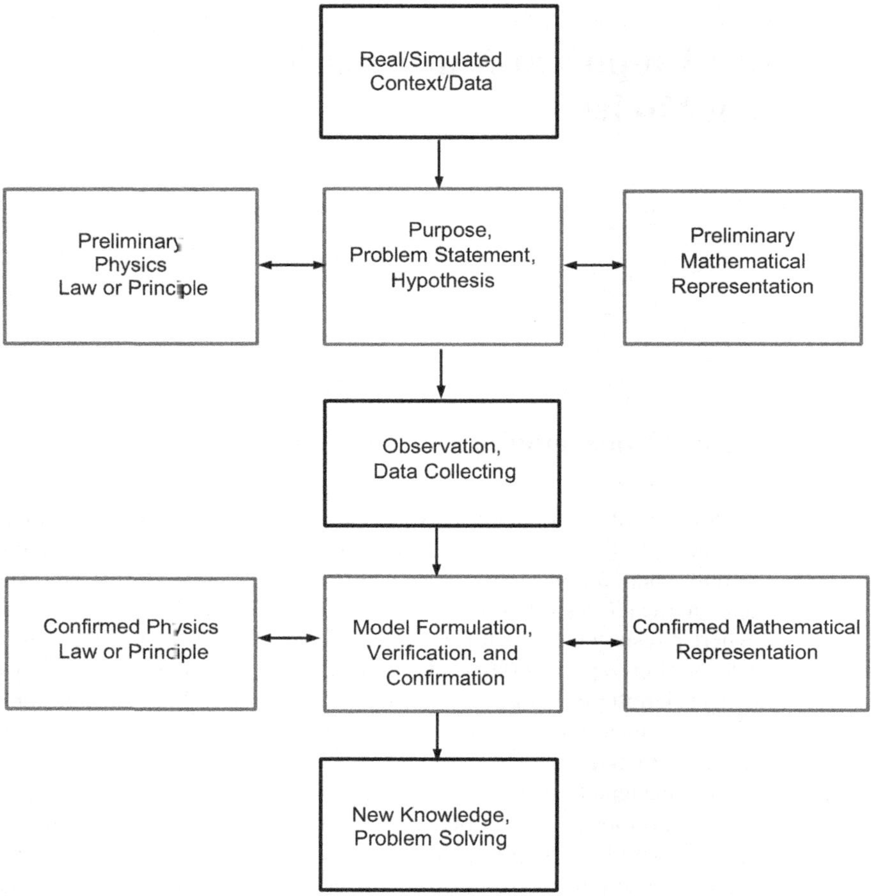

Fig. 4.1 Blended empirical-mathematical modeling scheme enhancing covariational reasoning

inquiry and will be assigned to be performed by the students. Students should be provided with instructional support that will guide them through these actions. The degree of support depends on the complexity of the phenomena and covariate relations that will emerge from the lab analyses.

The stem of the lab intertwines with two parallel chambers positioned on the left and right sides of the main actions that suggest merging math and physics constructs as they fit lab constraints. These constructs will fuse during investigations to produce an integrated inquiry. Such alignment is purposeful. Research showed that physics students have difficulties applying conceptual math knowledge to physics; thus, by explicitly highlighting this parallelism and mutual coherence, such merging will be more likely to occur.

The forms of representations that students will develop during the lab conducts can range from data tables and graphs to covariate structure formulation. A constant interplay between abstract mathematical symbolism and physics realities is to engage the learners in generating new knowledge. While the depth of the structural domain of mathematics is not elaborated explicitly, its baseline is drawn on the entanglements between the lab parameters embedded in the problem statement, hypothesis, and math knowledge that students acquired. The complexity of the algebraic structure brought to life during specific investigations will parallel with the traditional high school and undergraduate physics curricula. Due to its general purpose, the scheme can serve as a reference for using students' mathematical reasoning skills in other science branches as well.

The modeling process can be exercised using actual equipment, virtual simulations, or data provided by a table of values or other forms. It is suggested that if virtual simulations are used, they possess an exploratory character that allows for controlling variable parameters, data taking, and graphs plotting. It is also essential that the virtual context provides opportunities for verifications of the derived covariate relations. Simulations that meet all these conditions are often classified as computational (Davis et al., 2007). Students must be able to use their knowledge of physics and the language of mathematics to identify parameters of interest and classify them as dependent, independent. They might be encouraged to further categorize these variables as covariate parameters using more detailed scientific classification that is proposed in Sect. 5.2.

It is suggested that the instructor explain the lab's purpose and provide suggestions about the law or principle to be discovered or verified during the lab. The extend of the explanation will depend on the lab context and the grade level of the students. The instructor might introduce the experiment conduct and explain how to utilize the measuring devices to quantify parameters whether they are real or virtual. The lab introduction should be done, if possible, in a manner that will present it as a discovery adventure. The students always welcome a general lab presentation and how the experiment's parameters are set up.

4.3 Hypotheses as Learners' Proposed Theories

A problem statement typically formulated by the instructor is a form of unknown law and theory to discover upon lab completion. A catalyst of the lab conduct is formulating the hypothesis that reflects on the problem statement. The hypothesis is defined as the investigators' proposed theory supported by their prior knowledge explaining why the phenomenon under investigation behaves the way it does. Thus, depending on the nature of the investigation, hypotheses have the potential to be called law or theory (Simon, 2012). A reward for correctly hypothesized outcomes should be optimal, yet students undoubtedly welcome their hypotheses to be correct. Empowering the learners to formulate the hypotheses can be considered as passing the ownership to discover the underpinning of the experiment behavior to the students. Such action can inspire the students and increase their motivation to complete the lab. By formulating the hypotheses, students are placed in the position of being experts in understanding and merging physics and mathematics' laws and principles. Learners need time and *space* to develop, verify and revise their hypotheses if needed. While most typical lab activities are to verify already discovered scientific laws, there are plenty of opportunities for designing activities that enable new law formulation.

To ensure that the learners reflect on their mathematics knowledge and skills, they need to be unambiguously prompted to do so during the lab. This parallelism will be maintained throughout the lab by inserting questions that will require the knowledge merging. It is to assure that the switch from a verbally stated theory to a computational form of law is vital during the lab conduct.

4.4 Mainstream of the Inquiry and Its Confirmation

Conceptual questions precipitated during the lab conduct will enrich data taking and nurture the reasoning development that will merge into algebraic relationships. Research on problem-solving in physics stated that physics experts begin solving problems from a conceptual analysis of the physical situation then move into a more sophisticated mathematical analysis. In contrast, novices tend to start by selecting and manipulating equations (Larkin et al., 1980). To assure physics experts' inquiry order, proactive tasks will be embedded in the instructional support to direct the learners to take such path. In the proposed lab conduct design (Fig. 4.1), structural (conceptual) mathematics knowledge will constitute the main pilar of verifying algebraic structures that symbolically express the system's behavior and extend its hypothetical predictions.

4.5 Methods of Enacting Mathematical Structures

How will the algebraic relations be derived? The significant phases include identifying the parameters of interest, observing their mutual relationship, data gathering, graph plotting, and then zooming in on the graph possible algebraic representation. Depending on the system's complexity level, students might be offered supportive questions to select a correct algebraic expression for the data gathered. For example, to differentiate between selecting a linear or quadratic model, students might be prompted to discuss if the quantity of interest attains a maximum value or a constant rate of change. Questions of such prompts will help activate students' mathematical reasoning and identify such relations in the lab and problem-solving. Formulated mathematical models should correspond with what students predicted and observed in the experiment. Following the collections of such inferences, students will formulate an algebraic representation for the data. This phase of the inquiry might require revision of their hypothesized outcome and theory. There can be multiple supportive questions embeded in the verification phase to ensure that students reach the correct conclusions and revise and support them if needed.

4.6 Concluding Phases of the Learning Process

Students must be aware that revising, refining, and modifying their theories are necessary actions taken during scientific conduct. These actions enhance the quality of the investigations rather than diminish them. Revisions do not devaluate the investigator's knowledge and efforts, but they improve the scientific enterprise's validity. The revision process requires verification and establishing coherence between both scientific and mathematics reasoning outcomes. Students will be required to adhere to the scientific principles and their mathematical correctness when verifying formulated algebraic structures. While the reasoning process will develop an algebraic model of the phenomena, efforts will be made to extend the lab's inferences, bridge, and deploy these findings to problem solving that the students will encounter in their daily school practice. This extension is to have the students realize that the discovered and derived algebraic models are meaningful and helpful to succeed on assessment items. It is hoped that the inferences derived from the investigations will solidify in students' long-term memory as valuable impulses of knowledge that are ready to be retrieved when needed.

The proposed modeling scheme does not aspire to represent a universal theorem of modeling. It is rather to serve as a general guide to design labs in physics that aim at integrating physics inquiry with the structural domain of mathematics.

References

Davis, J. P., Eisenhardt, K. M., & Bingham, C. B. (2007). Developing theory through simulation methods. *Academy of Management Review, 32*, 480–499.

Larkin, J., McDermott, J., Simon, D. P., & Simon, H. A. (1980). Expert and novice performance in solving physics problems. *Science, 208*(4450), 1335–1342.

Simon, H. A. (2012). *Models of discovery: And other topics in the methods of science* (Vol. 54). Springer Science & Business Media.

Sokolowski, A. (2018). Formulating conceptual framework for multidisciplinary STEM modeling. In *Scientific inquiry in mathematics-theory and practice* (pp. 53–62). Springer.

Chapter 5
Covariational Reasoning – Theoretical Background

5.1 Quantities, Parameters, and Variables

Covariate reasoning can be exercised by observing selected parameters behavior and considering algebraic functions or formulas as structures depicting the behavior. Formulas are built up of varying and constant parameters that stem from a specific algebraic arrangement of quantities that warrant the experiment behavior. To realize their positions within the formula structures, the elementary elements need to be defined.

Quantities are the building blocks of physics constructs, laws, principles, and theorems. Quantities are inherent attributes of a specific object or parts of the system that support its numerical descriptions (Gilliard, 2020). For example, if the mass of the electron is concerned, the mass is an attribute of an electron. If the diameter of the Earth is concerned, the diameter represents an attribute of the Earth. These attributes are measurable, and their values possess certain magnitudes and units. Quantity can also be defined as someone's conceptualization of an object if the object has a measurable attribute (Fey, 1990). A variable represented by a symbol can be any one of a class of elements that may or may not be quantized. Following these definitions, quantities represented by parameters appear to be subsets to variables. How to differentiate between variables and parameters? A parameter in mathematics represents a constant quantity in the case under consideration but varies in other cases (Oxford University Press, 2015). Parameters in mathematics are usually denoted by alphabetical letters, a, b, c,etc., and the example (Eq. 5.1) of a quadratic function signifies these meanings.

$$g(x) = ax^2 + bx + c \tag{5.1}$$

The domain of this function is unrestricted and it can take any real value $x \in R$; however, there is a restriction on the parameter a in $g(x)$ to depict a quadratic

A. Sokolowski, *Understanding Physics Using Mathematical Reasoning*,
https://doi.org/10.1007/978-3-030-80205-9_5

function. The parameter cannot take a value of zero, $a \neq 0$, because the quadratic function would be reduced to a linear one $g(x) = bx + c$.

Following this analysis, a bridge to the quadratic formula (Eq. 5.2) that allows computing the x-intercepts of $g(x)$; can be made.

$$x = \frac{-b \pm \sqrt{b^2 - 4ac}}{2a} \tag{5.2}$$

The quadratic formula is dependent only on the values of its parameters, not on the independent variable x. In its covariate form, it can be expressed as $x(a,b,c) = \frac{-b \pm \sqrt{b^2 - 4ac}}{2a}$.

While this is apparent for teachers, for the students, it might not be. Is there a case in physics to parallel an application of this algebraic structure? Consider an object projected upward with an initial velocity v_1 from a height of y_1. The formula (Eq. 5.3) to find the vertical position of the object considered a function is

$$y(t) = y_1 + v_1 t - \frac{gt^2}{2} \tag{5.3}$$

A common task associated with function analysis is finding their horizontal intercepts, also called function zeros. The position function $y(t)$ can be used to find the time when the object lands on the ground by letting the vertical position of the object be 0, thus $0 = y(t)$. By rearranging the terms of Eq. (5.3), a quadratic equation in its standard form can be formulated $0 = gt^2 - 2v_1t - 2y_1$, which mirrors the case of finding the solutions of $ax^2 + bx + c = 0$ in mathematics by applying the formula (Eq. 5.2). Such mapping results in

$$t = \frac{v_1 \pm \sqrt{v_1^2 + 2y_1 g}}{g} \tag{5.4}$$

Since the formula produces positive and negative time values, one would consider only $t \geq 0$ as the acceptable solution(s). There is a specific condition for the parameters to produce real-time values; the value of the expression under the square root, also called the discriminant in mathematics must be greater than or equal to zero; $v_1^2 + 2y_1 g \geq 0$ otherwise, it would produce complex solutions. This condition does not surface physics resources, yet it presents itself as a compelling case for investigations. Furthermore, the formula to find the quadratic function x and y position of the vertex is:

$$x_v = \frac{-b}{2a} \text{and}\, y_v = f(x_v) \tag{5.5}$$

Referring to Eqs. (5.3) and (5.5) the time when the object reaches the maximum height can be calculated by $t_v = \frac{v_1}{g}$ and consequently, the maximum height by evaluating $y_{\max}(t_v) = y_1 + v_1(t_v) - \frac{g(t_v)^2}{2}$. The evaluation will result in $y_{\max}\left(\frac{v_1}{g}\right) = y_1 + v_1\left(\frac{v_1}{g}\right) - \frac{v_1^2}{2g} = y_1 + \frac{v_1^2}{2g}$. One will notice that $t = \frac{v_1}{g}$ can also result from formulating the velocity function and finding the time when the velocity of the object reaches a zero value. All these answers can be generated using kinematics formulas. However, offering students an alternative way can be an exciting experience of proving that interpretation of parameters in physics and mathematics leads to congruent conclusions. Using the properties of quadratic functions does not diminish the physics behind it; it instead equips students with more tools to solve physics problems and adds restrictions on the parameters to produce real output quantities.

When symbols of parameters are present in the function equation, the arrangement of variables defines a general rule for the family of functions. The values of parameters map up the structure to its particular case or graph. The idea can, to some degree, reflect on finding a general antiderivative and then generating a family of solutions by provided coordinate values and calculating the constant parameter c in the antiderivative. For instance, if $v(t) = \int(3t^2 + 2)dt$ is concerned, then the velocity functions are typically expressed as $v(t) = t^3 + 2t + c$. Thus, if an object accelerates with $a(t) = 3t^2 + 2$, its velocities are given by a family of functions $v(t) = t^3 + 2t + c$ that differ by the initial velocity denoted by c. The symbol c is called a constant in mathematics and physics resources; however, more precisely, it should perhaps be called a parameter whose value can be calculated due to the specificity of a particular case of motion. The symbol c represents a constant value when referred to a specific case motion, but it can take other values within the boundaries of the experiment setup. Parameters do not alter the general function form and the shapes of their graphs even though they constitute separate entities within function or equation. Parameters do alter the values of the function outputs or the values of the quantity of interest by the algebraic rules represented by the function structure. For example, $f(x) = x^2$ and $f(x) = 2x^2$ will produce different outputs for the same x values even though both equations will depict two quadratic functions whose general positions in the XY coordinates are similar. In a congruent fashion, the position function reflecting the formula $y(t) = y_1 + v_1 t - \frac{gt^2}{2}$ will generate similar graphs even though initial motion parameters like initial position and velocity might differ.

5.2 Formulas in Science and Mathematics

The term formula is widely used in sciences and mathematics as it manifests a mathematical model of real-world phenomena. The formula is defined in mathematics as an established rule of law expressed in algebraic symbols (Stewart, 2016). The building blocks of formulas are often called parameters, and they refer to specific quantities. Following this reasoning, one can conclude that quantities are being called parameters when embraced in algebraic formulas. The distinctions between variables and parameters are not often discussed in mathematics education. Research (Sokolowski, 2020) shows that even precalculus students have difficulties in differentiating between variables and parameters when a function or formulas are concerned. Exposing physics students to differentiating between these entities and developing their skills in this regard emerged as a task benefitting their general scientific and mathematical dispositions.

In science, a formula is a concise way of expressing a relation between quantities of a system using algebraic symbols. For example, $pV = nRT$ or $E = mc^2$. As such, a formula can entangle more than two quantities, and its complexity depends on the number of parameters that contribute to the phenomena behavior or investigators' research aim. Formulas are considered the most frequently used algebraic entities to solve or execute application problems in physics and other branches of science when quantification is concerned. While in physics, formulas help quantify specific instants of a phenomena behavior, in mathematics and especially in geometry, formulas are associated with an object's general properties and are used to calculate static assets such as volume, perimeter, area, or length, height, width, etc. At their vast applications, formulas in science and mathematics or other subjects resemble structures of algebraic functions. For instance, the formula for an area of a rectangle $A = ab$ can be expressed as $A(a) = ab$ or $A(b) = ab$ depending on what side is changing. Under that assumption, the formula can be called a linear function. While the right sides of the structures are unchanged, the left sides are different. Placing a parameter in parentheses directly, informs which parameter changes and which is constant.

Formulas are generally not considered as entities that can be sketched or analyzed from a function standpoint. Their applications are not embraced or restricted to specific values, although such restrictions are often needed. Consider, for example, the formula for the apparent frequency in the Doppler effect when the source is moving, $f'(v_s) = f_o\left(\dfrac{v \pm v_o}{v \pm v_s}\right)$. In this formula $f'(v_s)$ represents apparent frequency due to source movement, f_o the frequency produced by the source, v_s the speed of the source, v_o the speed of the observer and v the speed of the sound energy propagation in the given medium. When $v \pm v_s = 0$, the apparent frequency $f'(v_s)$ is undefined that creates a vertical asymptote on the graph of $f'(v_s)$, consequently, in the scientific sense, this situation produces a shock wave. The scientific interpretation of this case can be supported by taking sided limits of $f'(v_s)$ in the proximity of v_s which generates an exciting hypothetical analysis. Such analysis unpacks the scientific behavior of the phenomena not directly concludable from the formula notation. The hypothetical reasoning has its scientific support when the formula is interpreted as an algebraic function. Formulas are specific to a particular aspect of reality or *snapshot* of reality, but they can still be mapped up into their functional algebraic forms.

There are various ways of denoting algebraic functions and specifically underlying symbols of these entities that change. Such differentiating is not commonly applied when formulas are used, which when graphing or taking derivative or antiderivative is concerned might inevitably hide the variable nature of some parameters.

The notation that explicitly highlights the variable or variable parameter is called function notation and is placed in the parenthesis of the right side of the expression, e.g., $f(x)$ or $d(t)$. Function notation is typically not being applied in formulas to highlight the quantity that changes when presented in the general case. For example, in $U_G = -\frac{Gm_1m_2}{r}$, all the building blocks, objects' masses, and the separation between their center can change. The formula provides a general view of what parameters contribute to the change of the quantity of interest. However, in this regard, the formulas' scientific interpretations are reduced, perhaps therefore they are considered static entities. Considering formulas as static entities might disassociate these algebraic structures from the dynamism of the scientific processes they expose. By narrowing the usage of formulas to such applications, students miss the opportunities of perceiving formulas as covariate relations that can be sketched and whose behavior can be generalized. Consequently, students are unable to extract more general covariational relationships embedded within these structures. In the end, students lose the opportunity to develop conceptual skills for modeling function relationships in which the function output variable changes continuously in tandem with continuous changes in the input variable. Such reasoning abilities are essential for representing and interpreting the changing nature of a wide array of natural phenomena. The lack of such notation transcends to using these formulas in physics classrooms even when these entities are to be sketched. Moreover, the lack of highlighting the variable parameter by using function notation can dilute the process of finding the derivative or antiderivative. In calculus classes, the task of taking the derivative is embraced in the strict notation of the dependent and independent variable; $\frac{dv}{dt}$ or $v'(t)$ making these operations very explicit. By not emphasizing the notation that students learned in their math classes, the transitioning of math skills to physics might not be enabled.

5.3 Covariational Reasoning in Mathematics Education

Mathematical reasoning is not explicitly taught the same way mathematical concepts are being taught. It develops when students experience using numbers, quantities, variables, and relations between these quantities discussed earlier in Part II. Covariation reasoning (CR) appears as a more complex mental action that in this book will be considered a special category of mathematical reasoning. The relation of CR to mathematical reasoning is depicted in Fig. 5.1.

Mathematical reasoning generates a big picture of mathematical thinking methods. Covariational reasoning will be considered its more sophisticated derivative that includes attending to dynamic relations between selected parameters. The definitions and subsequent tasks of covariational reasoning as seen through the prisms

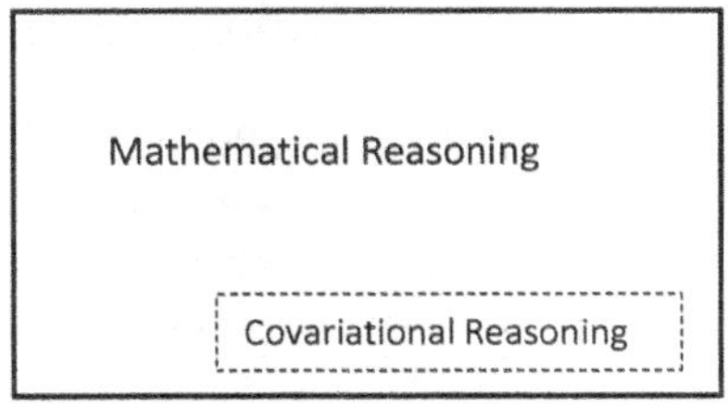

Fig. 5.1 Relation of covariational and mathematical reasoning

of physics and mathematics will be discussed, and an attempt to extract these meanings that will enhance the book's objective will be made.

While functions provide general frames for representing specific behavior concisely, formulas dressed up with contexts narrow applications of the representation to a specific part of reality. The mathematical mechanism of the formulas can be unearthed by analyzing covariation between the formula parameters. The meaning of the term covariation is closely related to variation. While *variation* refers to the diversity of one type of variable, covariation concerns the association of at least two different variables. Such association is also called a correspondence of the variations (Watson & Moritz, 2000). Covariational relations and reasoning are universally used in many branches of mathematics, and recently it caught the attention of physics education researchers. The definitions of a covariate parameter or variable are often augmented for the subject where they are being applied; thus, the definitions differ in mathematics, statistics, and science. At the same time, the definitions reflect on the subject domains and all exhibit certain commonalities. This section provides contemporary interpretations of covariation and suggestions for inducing such reasoning in physics teaching. A general definition of covariation that encompasses the meaning of this reasoning type found in mathematics research also emerged.

Covariation is a relation of mutual engagement of two quantities that becomes a gate to a broader cognitive tool of mathematical reasoning. Studies show that the function's perception as a process that accepts specific inputs and produces outputs is essential for a mature image of algebraic function (Johnson, 2015). This entanglement is essentially advanced in mathematics courses by embracing it in covariational reasoning. While mathematical reasoning and covariation are traditionally not taught, they can be developed through practicing and applying mathematical tools. Covariational reasoning as a theoretical construct appeared in mathematics in the late 1980s and early 1990s (Thompson & Carlson, 2017). Covariational relations strictly reflect on the dynamic behavior of a part of reality and can lead to formulating a symbolic representation of that reality.

There are several definitions of covariates and covariation that literature in mathematics and statistics education suggests. Carlson et al. (2002) defined covariation as a cognitive activity involving coordinating two varying quantities while attending to how they change with the relation to each other and advocated for emphasizing more covariate relations between functions and their derivatives to enhance students reasoning skills about function change. More explicitly, covariation is defined in statistics; Ott and Longnecker (2010) described a covariate as a variable related to the response variable (or parameter of interest) used to reduce the variability of the response variable. A covariate is also called a secondary variable (Moore, 2000). Following this definition, the independent variable is called primary. In contrast, covariate relationships would, according to this definition, be secondary to direct *cause and effect* relationships. These definitions place covariate relation as secondary to direct independent-dependent relation which does not correspond with the definitions of covariates by Thompson and Carlson (2017). Covariational variables as *secondary* variables are further classified as (a) *mediators* that support and help explain the relationship between the dependent and independent variables and (b) a *confounding* or *extraneous variable* whose effects are traditionally not welcomed in the experiments (Everitt, 2002). According to the definition, when the scientific experiment is concerned, confounding variables as extraneous variables that affect both the dependent and independent variables, which perhaps could transcend to factors that produce all kinds of errors during investigations related to accuracy and precision of data taking. To precisely estimate X's effect on Y, a researcher must prevent the effects of *cofounding* variables (Pearl, 2009). Cheng (1997) combined causal and covariation relations, stating that both must be induced from observable events. A causal relation between two parameters exists if the changes of the independent variable cause changes in the dependent variable. In this regard, the independent parameter is called the cause and the other the effect. Following Cheng's definition, the existence of casual and secondary covariational relationships must be observable and measurable. Johnson (2012) formulated three categories of covariation called *perspectives* which highlight the difference the investigator can perceive the relationships in a particular situation: (a) static covariation, (b) continuous dynamic, and (c) discrete dynamic. Research in mathematics education (Thompson et al., 1994) suggests that curriculum and instruction increase emphasis on advancing math students from a coordinated image of two variables to a coordinated image of their instantaneous rates of change. In Thompson's theory of quantitative reasoning, a person develops covariational reasoning when she/he envisions two quantities' values varying simultaneously. Following these descriptions covariation in conceptualizing individual quantities' values as varying and then conceptualizing two or more quantities as varying simultaneously.

In developing students' skills of coordinating changes of two quantities and describing the changes qualitatively and quantitatively, covariational appears as a precursor to understanding the concept of algebraic functions that is central to graphing and mathematizing phenomena using the rules and symbols of mathematics. Developing students' covariation reasoning abilities also concerns physics research communities; therefore, a brief perspective on how it is being used in physics courses will follow. Mixing all these perspectives will emerge as a proposal on how covariate relations will be considered and applied in the case studies in Part III of the volume.

5.4 Covariational Reasoning in Physics Education

Covariational reasoning and its effects on students' physics understanding is gaining momentum. However, research on covariational reasoning in physics research is still limiting (Panorkou & Germia, 2020). At the current stage, it is being developed conceptually and is similar in aim to its mathematical counterpart. For instance, the learners are provided with a bird's eye view or side view of a path of an object's undergoing periodic motion, and using these contexts; they are asked to sketch a graph of a distance of the object versus the total distance the object makes concerning a selected reference point (see Zimmerman et al., 2019). This relation can be described as the coordination of parameters viewed from different frames of reference. In other examples, students are asked to draw graphs of the object's speed versus position along know path of motion (see Monk, 1992). The cognitive value of such enterprises is unquestionable. However, a substantial focus on the conceptual development of covariation reasoning might unintentionally disjoin the accumulated students' skills from converting these prompts into concise algebraic representations. Constructing algebraic equations seems to be a task that students often encounter in their physics or other science courses; thus, not moving forward with formulating such for emerged covariance is making the efforts inadvertently missing their aim. Developing a conceptual understanding of covariates is one side of the coin; merging these covariations with algebraic structures and graphical representation of these functions is the other. Converting information to its graphical representation as a function requires identifying parameters that change (variables) and their classification as independent and dependent. Currently, these phrases are silent in physics education research. Can neglecting precise mathematical notations and parameter classifications in physics be one of the reasons for students' difficulties to effectively apply their math reasoning skills to graphing dynamic relations in physics? Findings of case studies described in Chaps. 12 and 13 provide evidence that a lack of precision of parameters labeling in physics humpers not only the interpretation of physics formulas but also jeopardizes graph sketching. Reforms in mathematics education (Hughes-Hallett, 2006) emphasize presenting functions in various forms; graphical, symbolic, verbal, and numerical to meet students' diverse learning styles and train them to transition between these representations. This didactical enterprise also inspired the design of the case studies clustered in Part III.

The term covariation already exists in physics education, and it broadly describes the casual relation between selected variables of the experiment. Covariation reasoning is applied to describe natural phenomena; thus, it involves actual quantities that are reimaged to parameters and assembled together in various algebraic relationships. These inquiries begin with identifying the parameters of interest and coordinating their intermediate relationships to their final product consisting of mental and symbolic images of their coordination. Following these processes, images of covariation are considered developmental phases that conclude with the formulation of the entire event. Applications of covariational reasoning are vast, ranging from their standard function relations and moving into the coordination of their rates of change or coordination of the functions of their components (e.g., two-dimensional motion).

In sum, while there are various definitions of covariations in mathematics, perceiving these special relations between two or more parameters will be used in this book and this definition will be a departing phase to propose a more detailed categorization of variables and parameters in physics teaching.

5.4.1 Viewing Phenomena as Covariations of Their Parameters

While there are several classifications of covariations in mathematics, literature does not provide equivalent classifications for sciences and, more specifically, physics, where such classification of the parameters is the priority. Thus, an attempt to synchronize these definitions is made. The classification was inspired by the synthesis of research and it is to generalize the meanings of various terms used and help understand the scientific underpinnings of the effects of parameters on the output of the investigation.

Physics is considered an experimental subject. Thus, investigations in physics are mainly experimental as opposed to just observational. The investigator actively manipulates certain variables associated with lab design, observes the parameters, takes measurements, and analyzes the data for potential covariate relationships. The quantities that the investigator manipulates can be called explanatory or independent covariates. Explanatory quantities generate specific changes in the quantity of interest, called also a response quantity, determined by the experiment design. There are usually other parameters involved in the experiment that affect the response quantity like confounding that generate measurement errors and mediating that link independent and dependent parameters. It is suggested, they these parameters are called secondary covariate quantities. Thus, secondary parameters will be further classified as confounding and mediators. The effects of the confounding quantities that fall into the basket of these factors that produce errors during data taking could be minimalized by, for example, employing more precise measuring devices or increasing the accuracy of data taking. Should such classifications be included during labs? Such classification does not seem to be necessary to assure that students focus on primary parameters; however, making students aware of all effects that contribute to the dependent parameter would enhance and broaden the perspectives on the experiment conducts. More detailed classification of parameters can be applied to formulas considered dynamic representations that can enrich students' perspectives on the techniques to solve problems.

This classification includes the terms, parameters, and variables as possessing similar meanings, although the term *variable parameter* appears to be more suitable. A *variable* in a mathematical sense carries the notion of a value (number) without a unit, and it commonly denotes an entity that can take various yet restricted values. Variables are classified as independent and dependent, thus using the term *variable parameter* will link the terminology used in both subjects.

Furthermore the term parameter embraces scientific interpretation on the variables, thus both adjectives; variable independent or dependent parameters seem to be suitable. In practice quantities that have the potential to change could be called e.g., variable force, variable mass, or variable resistance, and depending on the specifics of the experiment these would be further reclassified as dependent and independent. Thus to link the meanings, both terms will be used, and this is purposeful. Research (Sokolowski, 2020) shows that students do not have a clear image of how to differentiate between parameters and variables. Thus making these terms explicitly defined would help use them adequately.

Certain variations of each parameter are often present during experiments and each of the parameters (Fig. 5.2) will fit into that classification. This classification can be applied to enhance experiment descriptions and formula's understanding. Both venues are to help students master *cause-effect* reasoning and develop more profound insight into how phenomena behave also from their covariances viewpoints. For simplicity, the secondary covariates could also be perceived as indirectly affecting the experiment response variable (dependent parameter). The proposed classification does not include covariations of the rates of either the primary or secondary parameters. Such covariation can be derived using techniques of integration or differentiation and perhaps does not require an explicit categorization. Many physics students take a statistics course concurrently; thus, using the terms *mediating* and *confounding* in the same sense should help transition knowledge and enhance meanings. It is, though, not mandatory.

The section (Sect. 5.4.2) proposes how to use the classifications (Fig. 5.2) to support understanding of formulas' empirical-mathematical entanglements. Preliminary work on this section called for an attempt to categorize physics formulas due to how their covariates are executed which is also included. Section 5.4.3 attempts to provide an example of how to use classification (Fig. 5.2) while designing experiments.

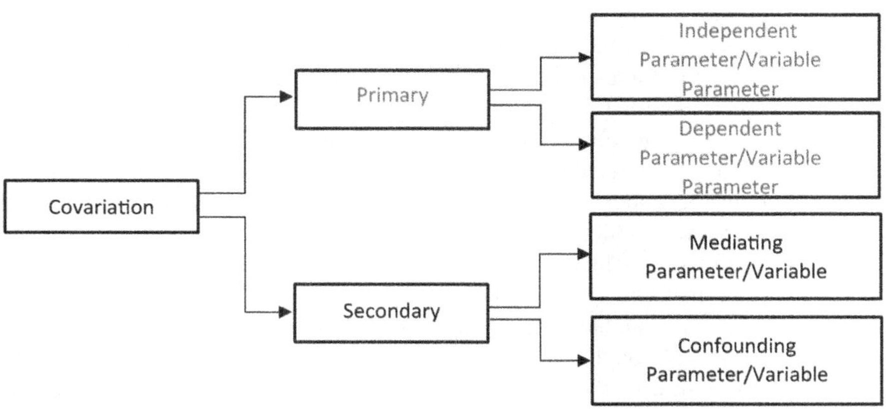

Fig. 5.2 Proposed classification of parameters in physics education

5.4.2 *Proposed Categories of Covariations Embedded in Physics Formulas*

Research shows that covariation can take various layers of complexity. Identifying primary and secondary covariates during experiments is one way of applying mathematical reasoning in sciences. Another could be examining the structures of various physics formulas and categorizing their structures considering the algebraic complexity of their covariations. The purpose of such an enterprise would be to offer students a more comprehensive view of describing natural phenomena algebraically. It is hypothesized that such discussions will support students' judgments in selecting correct algebraic representations to describe certain phenomena and let them search for their inventions. Physics formulas are built based on algebraic rules. The proposed classification of covariations reflects primarily on the underlying algebraic structures coupled with scientific interpretations. There are five categories identified and all are subject to possible revisions based on more exhaustive research. These categories do not include formulas presented as derivatives or antiderivatives for an earlier stated reason.

Type (1): Covariation involving a linear dependence
Examples of such variation can be formulas to calculate, for instance; the force of gravity acting on an object $F = mg$, wave equation $v = f\lambda$, or force exerted by stretched spring $F = -kx$. When embraced in typical functional notations, the formulas can take the following forms $F(m) = mg$, $v(f) = f\lambda$, or $F(x) = -kx$. Thus, when regarded as specifically formulated covariates, parameters of interest can be embraced in functional notation and positioned on the equations' left sides. The independent parameters are placed on the right side of these equations, however, their variable natures are denoted by placing their symbols in parenthesis. A rather stimulating discussion emerges when describing the effects of other parameters embedded within each of these formulas. From the mathematical standpoint in $F(m) = mg$, the parameter g representing gravitational field constant will be called a proportionality constant or the slope of the $F(m)$ function; from the proposed classification point of view, g can be considered a mediating parameter. As such, mediating parameter, g, also generates particular scientific insight. The gravitational field constant is, in this case, independent of the object's mass. It does, though, affect the object's weight or the force of gravity acting on it. Object's weight depends on g, not vice versa. As viewed through the prism of the proposed classification, all these constant quantities in; $F(m) = mg$, $v(f) = f\lambda$, and $F(x) = -kx$, thus g, λ, and k can be classified as *mediating* parameters. While g describes an external gravitational field, the wavelength depicts the medium that when coupled with vibration supplied by an external generator, affects the energy (wave) propagation speed. As the same mass has different weights depending on the magnitudes of the gravitational fields, the same energy vibration will move at different speeds depending on the medium that will alter the wave wavelengths. Referring to $F(x) = -kx$ spring constant k defines the property of the medium that is the spring under the extension labeled x. The

same extensions will require different forces depending on the mediating spring constant. While in algebra, a general form of a linear function is usually expressed as $f(x) = mx + b$, thus the slope listed as the first factor in the product mx, there seems to be no such rule in physics. For instance, in $F = mg$, the slope shows as the second factor, in $F(x) = -kx$, the slope k shows as the first factor. While this is not an obstacle in interpreting the formulas, pointing this lack of consistency to students will make them aware of the flexibility they need to exhibit while interpreting the physics formulas

Formulas can consist of more than one parameter, e.g., $F_B = \sigma Vg$. A general interpretation of such formulas would gear toward considering one parameter as the mediating, in this case, g, and the density of the fluid σ along with fluid displaced volume V as constituting independent parameters. Such classification would perhaps depend solely on the specific experimental setup or problem to analyze.

For example, one of these parameters σ or V can also be considered a mediating one, and the other labeled as a variable one. Certainly, referring to specific objectives of the experiments will make such classifications more explicit. All these quantities of interest increase in magnitude when the independent parameter increases, and the changes are proportional to mediating parameters. Such discussions seem to bring new insight into these formulas and their graphical interpretations. While the slopes of each of these formulas visualize mediating parameters, the mediating parameters do all exist independently from the external actions. These actions can be used to visualize and quantify all these mediators mathematically. There are certainly more physics formulas of Type 1 structures.

Type (2): Covariation involving a rational dependence

Examples of such covariations can be $a = \frac{Gm}{d^2}$ or $R = \frac{\sigma A}{l}$. When converted to functional dependencies, they can take the following forms $a(d) = \frac{Gm}{d^2}$ or $R(l) = \frac{\sigma A}{l}$. The effects of the independent parameters on the parameter of interest depend on a selected parameter's algebraic position within the symbolic formula representation. If the independent parameter takes numerical values, its variation is described by sequencing it in increasing order, thus from lower to larger magnitudes, and respectively the effects are justified due to covariation. For example, the dependence of the gravitational field intensity from the object's mass or the distance from its center. This covariation typically takes two different forms as $a(m) = \frac{Gm}{d^2}$ which is linear or $a(d) = \frac{Gm}{d^2}$ which is rational. In special cases, the magnitude of the intensity can be affected by a varying mass m and varying distance d which could be denoted as $a(m,d)$. All examples describe the properties of gravitational field produced by the mass m. Can G, the universal gravitational constant, be considered a mediator of the phenomenon? This is an open question. G undoubtedly supports quantification of the phenomena, but it is a coefficient in this formula that makes the formula universal. The other formula, $R = \frac{\sigma A}{l}$, describes the quantification of a conductor resistance considering its geometrical parameters and resistivity.

When considered $R(l) = \frac{\sigma A}{l}$, the formula resembles a rational function which tells that the resistance decreases as the length of the wire increases. When considering the cross-sectional area as the independent parameter, $R(A) = \frac{\sigma A}{l}$, the resistance R depends linearly on the area. From students' points of view, such flexibility in interpretations might not be apparent. While standard covariations of two variable parameters cannot be sketched in the traditional XY plane e.g. $a(m,d)$ or $R(l,A)$, physics provides many examples when students must find an expression for a new value when two independent parameters change. While such algebraic manipulations seem to be straightforward, the mathematics curriculum does not provide many opportunities for practicing such skills; therefore, to enhance knowledge transfer, physics teachers' guidance is welcomed and appreciated by the students.

Type (3): Linking two or more dependent parameters

Linking the variation of two dependent parameters can also lead to learning new information about the phenomena. A good example to discuss can be two-dimensional motion. While the causes of motion in the vertical direction do not affect the horizontal motion, coordinating, for example, vertical and horizontal positions, thus $x(t)$ and $y(t)$ can emerge as the path of the object's motion. This is possible because the independent parameter, t, is shared for both functions, and evaluation of both of these functions $x(t)$ and $y(t)$ generates position coordinates (x,y) and the table of values for $y(x)$. It is to note that in the final form, the *path of motion* does not contain that parameter. Type 3 covariation can also include finding component velocity and acceleration functions; $v_x(t)$ and $v_y(t)$ along with $a_x(t)$ and $a_y(t)$ that in a similar manner to position functions can be used to finding resultant velocity and acceleration functions.

Another example of using such covariation could be analyzing the temperature in two different scales $C(t)$ and $K(t)$. For example, a scientist measures the temperature of a substance at selected time instants using two thermometers with two different scales Celsius and Kelvin, and creates $C(t)$ and $K(t)$ functions. If time is considered the independent parameter, the temperature will be called the dependent. The thermometers will measure the temperature independently, producing different values because of the different scales. Finding an algebraic function that shows a covariation of the temperatures can be used to derive conversion equations, either $C(K)$ or $K(C)$. When mediating parameters are sought in this enterprise, one could pinpoint the specific heat capacity of the substance used in the experiment, affecting the temperature increase rate. However, the heat capacity will not affect the general conversion formulas $C(K)$ or $K(C)$ because measurements on both will be altered on both thermometers at the same rates dictated by the substance heat capacity.

Type (4): Covariation of multiple parameters within a system

This category would involve systems of objects or one object simultaneously examined using more than one physical concept. These structures are often called laws, e.g., the law of conservation of momentum, law of conservation of mechanical energy, gas law. The parameters constituting these structures can also be embraced in analyzing their dependent, independent, or mediating effects. Since laws encompass usually more objects and parameters, there is greater flexibility in classifying the parameters. Let us consider the law of conservation of mechanical energy with kinetic and potential energies involved. For instance, if an object of mass 2 kg that moves in two dimensions possesses only kinetic and gravitational energies and no frictional effects are present, then at any two selected time instants or positions, the total mechanical energy remains the same, and thus $2gh_1 + \frac{2v_1^2}{2} = 2gh_2 + \frac{2v_2^2}{2}$ holds to be true. The law can be converted to a more direct covariate expression upon assigning the dependent and independent parameters and its analysis enriched by considering covariations of the system parameters. Suppose that the total initial mechanical energy of the object is 200 J, and one is interested in attending to the velocity function expressed in terms of the variable height to which the object can rise. Following this condition, $200 = 2gh2 + v_2^2$ and solving for $v_2(h_2) = \sqrt{200 - 2gh_2}$ provides means of calculating the magnitude of the velocity at any height h_2 along the motion path and sketch it in respective coordinates. The independent variable in this expression is the vertical position of the object above a reference level labeled h_2, and g is a mediator. Since a new algebraic representation was derived, there can be a new analysis imposed on it. The radical sign generates certain restrictions on the on variable heigh h_2 such that $200 - 2gh_2 \geq 0$, which leads to $h_2 '' \frac{100}{g}$. This restriction tells that with the conditions provided, the maximum height reached by the object is about 10 m. Students can extend their hypothetical reasoning and discuss the effects of the mediating parameter; the gravitational intensity on the object's maximum height. The analysis would take a similar aim, resulting in a different final product if the dependent parameter is the object height h_2 instead of speed labelled v_2. Laws involving systems of objects provide more opportunities to generate algebraic structures because of the diversity and richness of the parameters from which they are built.

Type (5): Scaffolding covariation

This covariation is not surfacing physics formulas, but it is useful when composites of two functions are concerned or when the independent parameter depends yet on another formula. Algebraically, such covariations are expressed as $y = f(g(x))$ where $g(x)$ is considered an inner function of $f(x)$. While problem-solving typically does not require creating composite formulas, such algebraic operations can support graphing and formulas deriving from different, perhaps more insightful perspectives. For example, the gravitational potential energy of an object is calculated using $U(y) = mgy$, where y represents the position of the object above or below the established frame of reference. If the object's position changes according to $h(t) = 2t + 4$, then the object's gravitational potential energy can be expressed in terms of time t thus $U(h(t)) = U(t) = mg(2t + 4)$. The units stemming of the new formula must still be *joules*, but the independent variable is no longer the object's position but it is the time, t. Such representations of the gravitational potential energy add dynamism to the formulas and allow to predict and verify the energy for a range of time values.

Another example could be $K = \frac{1}{2} I\omega^2$. If the variable parameter is the angular velocity ω, then the formula can be expressed as a function of that parameter, thus $K(\omega) = \frac{1}{2} I\omega^2$. If the behavior of the angular velocity function is given, e.g., $\omega(t) = \omega_0 + \alpha t$, then $K(t) = \frac{1}{2} I(\omega_0 + \alpha t)^2$.

Students study the techniques of finding composite functions in a typical high school math curriculum, thus, transferring these skills to physics should be possible. Another example of this category can be the formula for pendulum period $T = 2\pi\sqrt{\frac{l}{g}}$, which in school practice is considered as $T(l) = 2\pi\sqrt{\frac{l}{g}}$. This formula already embeds an inner function $\frac{l}{g}$ which likely has no special physical meaning.

Discussions about finding commonalities in physics formulas provide a different perspective on these entities and open up a gate to broaden student's perspectives on the universality of these structures and their fit to describing scientific phenomena. Consequently demystifying the algebraic rules that the formulas are build from should also help students set up equations to solve problems in physics.

The categories suggested in this section are not exhaustive and they can take more aims. It seems that as long as they enrich students' views on physics methods and help with understanding, they serve their purposes.

5.4.3 Discussing Covariations of Parameters in Experiments

This section is an attempt to provide examples of classifying parameters when experiments or laboratory activities are concerned. Experiments provide an opportunity to include confounding parameters, thus involved parameters can be classified as dependent, independent, mediating, and confounding. Any experiments, especially at their designing stage, can involve several parameters, yet not all are necessarily used to achieve the lab goal. Selecting and classifying variables will be followed by selecting respective measuring devices needed for the lab completion; for example, a sensor to measure the object's initial speed, tape to measure length, etc. The discussion that follows is preliminary, and indeed, other venues can emerge due to range of available measuring devices, equipment, or creativity of the lab designer.

Example 1: Suppose that an object is projected straight up with different velocities, and an algebraic relation between the height and the initial velocity is sought.

The response variable is the height, and the input variable is the object's initial velocity. The height reached by the object depends on the gravitational field and the time. These two parameters will be called *mediators*. They affect the height, but they will not be included in the analysis. There is also a force of air resistance that decreases the object's height because of its effect on the object's net force. The air resistance can be called a confounding variable. This experiment could aim differently; it could investigate the time of motion in the object's initial velocity function. Such lab profile requires reclassifying the parameters. The response variable will be now the time, and the independent parameter the initial velocity. The intensity of the gravitational field would retain to be called a mediator and the air resistance as the confounding parameter.

Example 2: Investigating the coefficient of kinetic friction between the wooden block and the floor by pushing an object along the floor so that it moves a definite distance and it stops.

Suppose that the available measuring devices are tape to measure the length and a stopwatch. The process of Identifying and classifying the parameters involved in computations of the coefficient of friction is not that straightforward due to how the parameter of interest (response variable) relates to the distance covered and time of motion. While $F_f = \mu F_N$, the parameter of interest μ is not directly related to kinematics of the motion that could be measured using available devices. Thus, Newton's law $F_{Net} = \; - F_f$ must be used to find the value of the left side of the equation. Considering $F_{Net} = ma$ and $x = \frac{at^2}{2}$ mediating formulas, the coefficient of friction can be found. While unfolding the covariates, more parameters are being included, and all enrich the process. Is it necessary to classify all the parameters? Such a task is interesting and it would enhance the connections between all the parameters. By realizing the covariate links between the quantities, students would have an opportunity to envision a holistic view of the process, understand its building blocks that in return will help them remember it.

Example 3: Measuring the effects of differentiating the potential difference on the current flowing a resistor.

This experiment would resemble Type 1 covariation, with the independent parameter the potential difference and the current the dependent variable. The internal properties of the resistors will emerge as the mediating parameter representing the electrical resistance. An extension of such lab could be having two such resistors and investigating their equivalent resistance when connected in series and parallel. The connection types resulting in different equivalent resistance in this case will mediate with the value of the electric current. Thus it can be categorized as yet another mediating parameter that graphically is represented as a slope in potential difference versus the current graph.

Classifying parameters as dependent, independent, or mediating requires a more profound analysis when experiments are concerned. Bringing forth a dynamic nature of the formulas should enable students to perceive these mathematical entities as functions that depict the experiment behavior. Being provided with various lab contexts will let the students' exhibit flexibility in assigning quantities as variables that should also influence their problem-solving skills. It is hoped that these preliminary classifications provide departing points for a more detailed analysis that would extend generalizations of the phenomenon behavior stemming from either experiments or formulas.

5.5 Limiting Case Analysis

Limiting case analysis has already been surfaced in this book content, and it will be used more extensively in Chaps. 6 and 9. This section provides more algebraic details about limits evaluation in the context of physics. Using the idea of limits allows for extending both predictions and explanations of physical phenomena. According to Redish (2017), these types of analyses should be essential goals of teaching physics because they allow for generalizing inferences by evaluating or estimating values of algebraic formulas considered covariate functions. In practice, such a transition from formulas to functions provides opportunities for deriving valid conclusions for cases when direct laboratory measurements or formula evaluations are not possible. While using limits is common for scientists, the idea of applying limits in school practice is not visible, and testing students' ability in this area is also rare (Sokolowski, 2018). Historically, the idea of limits was developed by Archimedes of Syracuse in the third century BC. The theory of limits was independently developed by Newton and Leibnitz in the seventeenth century (Burton, 2007), and it sets the foundation of modern calculus. The concept of limits has been used to derive the concept of derivative and accumulation as well as other fundamental ideas in calculus. Due to its diverse applications and being described in various, often very abstract ways, its understanding by the students of mathematics causes challenges (Vinner, 2002), therefore presenting it to students from the application point of view should help with understanding. To use limits in physics, its definition will be augmented so that students see its relevance, usefulness, and help to understand physics ideas. The following definition of limits is suggested:

> *Limits are mathematical tools that help evaluate formulas expressed as covariate algebraic representations in situations when a direct evaluation is not possible.*

As such, limits are induced to covariate dynamic representations and help explicate the nature of these covariates. Being able to pinpoint the independent parameter and its algebraic relation with the dependent one appears the first step toward understating applications of limits in science. Consequently, to understand why limits of formulas can be taken, students have to perceive formulas as dynamic algebraic relations that have their equivalent representations in typical XY representations. Limits are closely related to functions diverse expressions, symbolic, graphical, or tables of values with explicitly defined variables. To help students link techniques of taking limits that they learn in mathematics with inducing these techniques to physics formulas, physics formulas need to be reimagined, and the parameters need to be reclassified as constant or variable. The pathway to structuralize the induction of the limiting case analysis to analyze natural phenomena is illustrated in Fig. 5.3.

There will be generally two types of limits applied in this book (a) when the values of the independent parameter are getting very large and (b) when the values of the independent parameter are minimal and close to a zero magnitude. Both methods result in producing approximated values of the dependent parameter that can be considered by the definition of limits its best approximation. The discussion

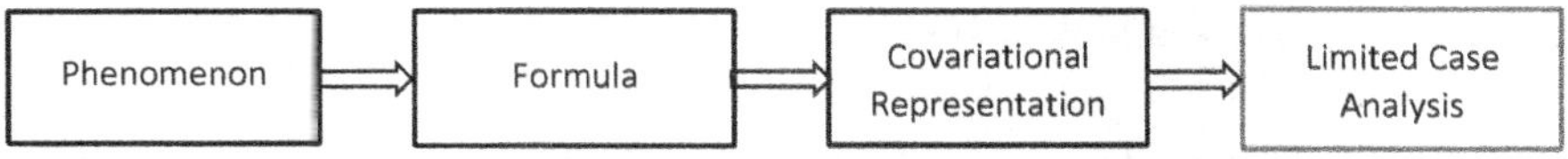

Fig. 5.3 Schematic representation of the theoretical framework when limiting case analysis is used

of the techniques of taking limits for very large and very small values of the independent variable will be conducted using an electric circuit with two resistors connected in parallel as a model. Presenting these ideas in a context should enhance the understanding of the result. A complete discussion of this topic is presented in The Physics Teacher (Sokolowski, 2019a).

5.5.1 *Evaluating Limits when the Variable Parameter Is Getting Very Large;* $x \to \infty$

Suppose that one analyzes a parallel connection of two resistors such that one resistor has a constant resistance, e.g., $R_1 = 10\ \Omega$ and the other has a variable resistance. Note that one of the resistors needs to have a constant value to represent a one-variable function. Alternatively, the value of one resistor could be expressed in terms of the other.

The specific task is to determine the equivalent or total resistance when the resistance of the variable resistor is getting a substantial and infinite value. First, we need to identify a correct formula related to finding equivalent resistance of two resistors connected in parallel that is:

$$\frac{1}{R_T} = \frac{1}{R_1} + \frac{1}{R_2} \tag{5.6}$$

The formula (Eq. 5.6) does not explicitly display a covariate representation. Thus, applying the techniques of taking limits using students' math knowledge is not recommended, although possible using Eq. (5.6). Furthermore, Eq. (5.6) does not allow for visualizing the idea by sketching the function and its approximated value at infinity; thus, relying on the rules of mathematics and rearranging the formula is needed To apply the techniques of taking limits that students study in mathematics and visualize the result, the formula needs to be reimaged to a rational functional form. It is convenient to label the variable resistor as x because it will resonate with forms that students analyzed in their mathematics courses.

$$R_T(x) = \frac{10x}{x+10} \tag{5.7}$$

While both forms (Eqs. 5.6 and 5.7) can lead the students to reach similar final answers, students can easily recognize the function equation in Eq. (5.7) as a rational function and the appropriate technique to taking the limit. The limit must be induced because substituting infinity for x results in an undetermined form $\frac{o}{o}$ whose value can not be approximated. While the phrase *take limit* dictates specific tasks in mathematics, in physics populated by contexts, the phrase might be modified to say, for example, determine the value of the equivalent resistance of the circuit for a considerable significant value of the resistor labeled R_2. How students denote the algorithm of taking function limits in mathematics? Suppose that $f(x) = \frac{10x}{x+10}$, then the process is typically initiated by embracing the function expression in the limit notation $\lim_{x\to\infty} \frac{10x}{x+10}$. Realizing that the function is rational, one would consider only its dominant terms as the terms contributing to the function value for very large x values; $\lim_{x\to\infty} \frac{10x}{x+10} = \lim_{x\to\infty} \frac{10x}{x} = 10$. One can notice that the notation does not explicate the critical information that the value of the limit represents the approximation of the function $f(x)$ for substantial x values. While it is expected that students realize this result, and perhaps they do, omitting the notation might cause some uncertainties. To make the association of the limit output to function (formula) value more apparent this process can be explicitly labeled. Referring back to our example; to enhance this understanding, the following notation is suggested:

$$R_T(x \to \infty) = \lim_{x\to\infty} \frac{10x}{x+10} = \lim_{x\to\infty} \frac{10x}{x} = 10\Omega \tag{5.8}$$

Including the function notation on the left side with the symbol of the independent parameter explicates the results making it clear to interpret. By using $x\to\infty$, not $R(\infty)$, the rules of mathematics are also respected. One would read Eq. (5.8) following; the value of the equivalent resistor of the circuit as the resistance of R_2 is very large is 10 Ω. There is another nuance worthy of discussing; the interpretation of the limit value in the context of the function. While the equal sign is being used to denote the limit value, the value of 10 Ω approximates the equivalent resistance in these circumstances. It is not the accurate value of the total resistance. This interpretation can be visualized (see Fig. 5.4) by sketching the limit value (the dotted red line) and the Eq. (5.7). Thus the result of taking limit of Eq. (5.8) represents the horizontal asymptote of $R_T(x) = \frac{10x}{x+10}$ and the value of the limit. Horizontal asymptotes are described as the boundaries of function values for very large (positive or negative) values of independent variables. Asymptotes are not considered parts of functions, but they well approximate their values.

Graphical representations are easier accessible by students than symbolic; thus, illustrating the idea not only conveniently shows the limiting value of the equivalent resistance, but also allows for tracing the dynamism of the total resistance when the resistance of one resistor increases.

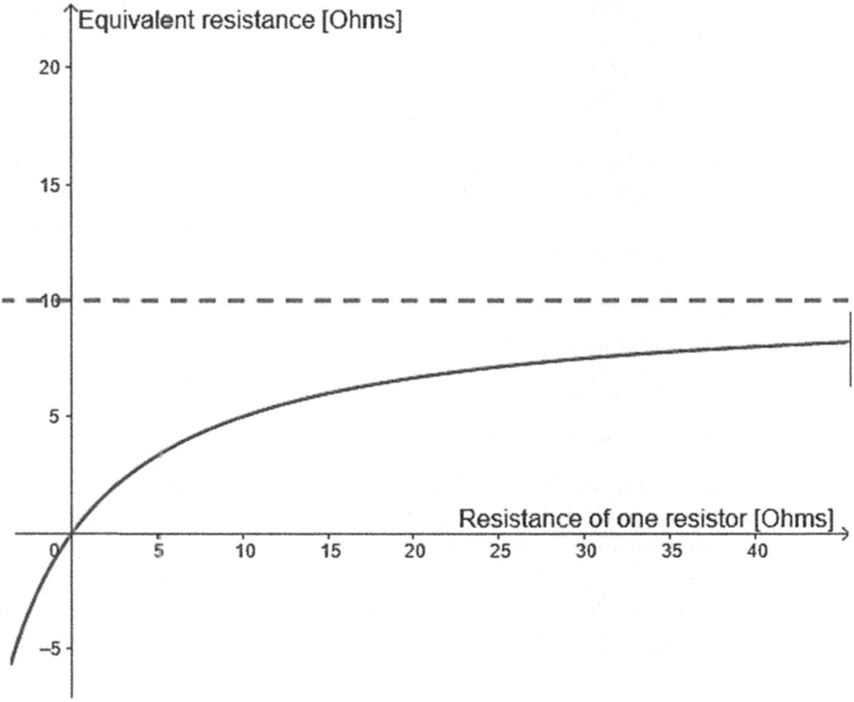

Fig. 5.4 Graph of $R_T(x) = \frac{10x}{x+10}$ restricted to $x > 0$ showing the value of the total resistance when $x \to \infty$

5.5.2 *Evaluating Limits when the Variable Parameter Is Close to a Specific Value;* x→a

Evaluating function limits for a very small value of the independent variable is rare in physics education as well, yet its output often provides even more interesting facts about the phenomena behavior than limits taken for large value of the independent parameter. Limit when $x \to a$ is often rewritten as sided limits, and it carries two independent operations as $x \to a^+$ and $x \to a^-$ that tell the direction of function evaluation; respectively from the right or the left side of of a. The technique of evaluating such limits requires a more detailed analysis. In contrast, applying it to evaluate the phenomena behavior can be very rewarding.

Let us consider again the parallel connection discussed in Sect. 5.5.1 and focus this time on analyzing the value of the total electric current drawn from the battery when the resistor R_2 is reaching a value close to zero. Let us assume that the potential difference in the battery is 9 V. The electric current depends on the total resistance of the circuit by $I = \frac{V}{R_T} = \frac{9V}{R_T}$. Since the expression for R_T has been formulated, then $I(R_T)$ provides means for formulating the electric current function.

$$I\left(R_T\left(x\right)\right)=I\left(x\right)=\frac{9}{\dfrac{10x}{x+10}}=\frac{9x+90}{10x} \tag{5.9}$$

This composite function (scaffolding covariate) represents another rational function whose graph is illustrated in Fig. 5.5.

When $x = 0\ \Omega$, the function is undefined, consequently one can say the electric current has an undefined value, but the idea of a sided limit can be induced to approximate the value by taking the right-sided limit of the function $I(x \to 0^+) = \lim_{x\to 0^+} I\left(x\right)=\lim_{x\to 0^+}\frac{9\left(0.001\right)+90}{10\left(0.001\right)}=\infty$. This result proves that the electric current could take an infinite value when the resistance R_2 is close to zero ($x \to 0\ \Omega$). The infinite right-sided limit of the *current* function suggests a boundary for the value that is illustrated in Fig. 5.5 as a vertical asymptote $x = 0\ \Omega$. The vertical asymptote is depicted by the red vertical dashed line positioned along the y-axis. To expand on

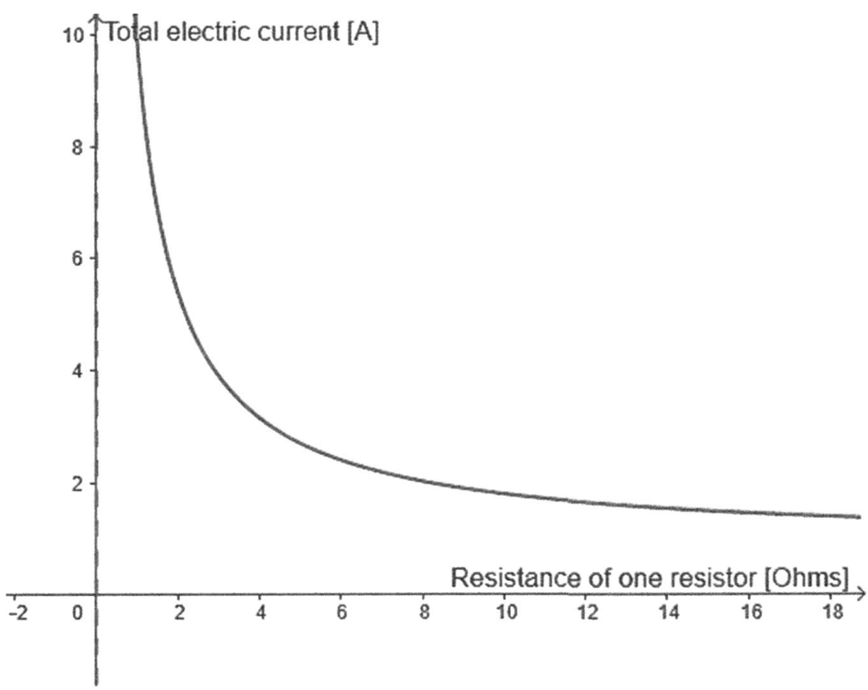

Fig. 5.5 Electric current expressed in the function of the variable resistance $I\left(x\right)=\frac{9x+90}{10x}$

the underlying scientific behavior, one can conclude that when the resistance of one resistor in a parallel connection takes the value of zero, then almost all electrons take the path with the zeroth resistance to flow creating a subsential electric current. It is to note that the left-sided limit $x \rightarrow 0^-$ was not be evaluated because the analysis is valid only when $x > 0$.

The physics curriculum provides multiple possibilities for inducing limits and extend the scientific inquiry of natural phenomena beyond classroom experiments and textbook problems. Students find it exciting to learn that complex calculus concepts have applications, and these applications extend their knowledge about nature.

5.5.3 Is Limiting Case Analysis Really "Limiting"?

In commonly used calculus textbooks, problems requiring students to take limits are usually phrased like: *state the value of each quantity*, *state the value of the limit* or *find the limit* (Stewart, 2016). Thus, a phrase *applying limiting analysis* is not being used in mathematics education as a standard task dictating limit application; instead, students are prompted to use function evaluation techniques and take function limits. Taking the limit of a function appears as another way of function evaluation rather than a different way of analyzing function.

While termed *limiting case analysis* in physics education such task command does not seem to be necessary. The term *limiting* might suggest that the independent or dependent parameter's values are small or limited to specific fixed magnitudes. While this can be true in some cases, in most cases, limits are applied to extend the inquiry for substantial values of the independent parameter. The phrase *limiting case analysis* will be used in this book to allow transitioning of terminology between physics and mathematics, however, it will not be emphasized, and its alternative prompt: *evaluction of the covariational expression* by taking a limit, will be employed. Generating a new mathematics language to use in physics might inadvertently create a notion that the technique of applying limits is restricted only to a domain of physics that in fact does not reflect the range of limits applicability. In short, using the terminology that students use in mathematics is seen as a factor helping students with transferring their math skills to physics even when applying sophisticated apparatus. However, while in their mathematics courses, questions that ask students to take function limits are explicitly verbalized or symbolically expressed, there seems to be no agreement on how to verbalize questions to have students realize that limits are to be taken when applications are concerned in physics. Will a direct command of taking the limit of a functional formula be reasonable? Or should instead, a phrase about the progression of the independent parameter values be clearly described suggesting students take limits to answer questions? There is a room to address this nuance and find the most optimal phrases that will satisfy both mathematics rigor and scientific interpretation.

References

Burton, D. (2007). *The history of mathematics: An introduction*. McGraw-Hill.

Carlson, M., Jacobs, S., Coe, E., Larsen, S., & Hsu, E. (2002). Applying covariational reasoning while modeling dynamic events: A framework and a study. *Journal for Research in Mathematics Education, 33*, 352–378.

Cheng, P. (1997). From covariation to causation: A causal power theory. *Psychological Review, 104*(2), 367–405.

Everitt, B. S. (2002). *The Cambridge dictionary of statistics* (2nd ed.). Cambridge University Press.

Fey, J. T. (1990). Quantity. In *On the shoulders of giants: New approaches to numeracy* (pp. 61–94). National Academy Press.

Gilliard, R. P. (2020). Fundamental units and constants of physics. *Advanced Studies in Theoretical Physics, 14*(4), 209–217.

Hughes-Hallett, D. (2006). What have we learned from calculus reform? The road to conceptual understanding. In N. Hastings (Ed.), *Rethinking the courses below calculus*. Mathematics Association of America.

Johnson, H. L. (2012). Reasoning about variation in the intensity of change in covarying quantities involved in the rate of change. *The Journal of Mathematical Behavior, 31*(3), 313–330.

Johnson, H. L. (2015). Secondary students' quantification of ratio and rate: A framework for reasoning about changes in covarying quantities. *Mathematical Thinking and Learning, 17*(1), 64–90.

Monk, S. (1992). Students' understanding of a function given by a physical model. In G. Harel & E. Dubinsky (Eds.), *The concept of function: Aspects of epistemology and pedagogy* (MAA Notes) (Vol. 25, pp. 175–193).

Moore, T. (2000). *Teaching statistics: Resources for undergraduate instructors*. Mathematical Association of America.

Ott, R. L., & Longnecker, M. T. (2010). *An introduction to statistical methods and data analysis* (6th ed.). Brooks/Cole.

Oxford University Press. (2015). Parameter. In *Oxford dictionary of science* (7th ed., p. 586). Oxford University Press.

Panorkou, N., & Germia, E. F. (2020). Integrating math and science content through covariational reasoning: The case of gravity. *Mathematical Thinking and Learning*, 1–26. https://doi.org/10.1080/10986065.2020.1814977

Pearl, J. (2009). Simpson's paradox, confounding, and collapsibility. In *Causality: Models, reasoning and inference* (2nd ed.). Cambridge University Press.

Redish, E. F. (2017). Analyzing the competency of mathematical modelling in physics. In *Key competencies in physics teaching and learning* (pp. 25–40). Springer.

Sokolowski, A. (2018). Modeling acceleration of a system of two objects using the concept of limits. *The Physics Teacher, 56*(1), 40–42.

Sokolowski, A. (2019a). Applying structural mathematics in physics: Case of parallel connection. *The Physics Teacher, 57*(9), 627–629.

Sokolowski, A. (2020). Like terms in algebra-is the current definition adequate? Proposal for an instructional unit for high school students. *Australian Mathematics Education Journal, 2*(2), 13.

Stewart, J. (2016). *Calculus: Single variable calculus* (8th ed.). Brooks/Cole.

Thompson, P. W., & Carlson, M. P. (2017). Variation, covariation, and functions: Foundational ways of thinking mathematically. *Compendium for Research in Mathematics Education*, 421–456.

Thompson, P. W., Hatfield, N. J., Yoon, H., Joshua, S., & Byerley, C. (2017). Covariational reasoning among US and South Korean secondary mathematics teachers. *The Journal of Mathematical Behavior, 48*, 95–111.

Thompson, P. W., Philipp, R., & Boyd, B. (1994). Calculational and conceptual orientations in teaching mathematics. In *1994 Yearbook of the NCTM* (pp. 79–92). NCTM.
Vinner, S. (2002). The role of definitions in the teaching and learning of mathematics. In *Advanced mathematical thinking* (pp. 65–81). Springer.
Watson, J. M., & Moritz, J. B. (2000). The longitudinal development of understanding of average. *Mathematical Thinking and Learning, 2*(1-2), 11–50.
Zimmerman, C., Olsho, A., Loverude, M., Boudreaux, A., Smith, T., & Brahmia, S. W. (2019). Towards understanding and characterizing expert covariational reasoning in physics. *arXiv preprint arXiv:1911 01598*.

Part III
From Research to Practice

Chapter 6
Extending the Inquiry of Newton's Second Law by Using Limiting Case Analysis

6.1 Limits - Tools for Extending Scientific Inquiry

This research aims to present an example of applying limits to extend the inquiry in physics. While most current developments in this area take the stance that students' mathematical reasoning can be enriched by improving students' conceptual understanding of physics formulas, this study aims differently. It aims to extract phenomena behavior from formulas considered functional covariation relations. What prompted such aim? Research (Sokolowski 2020) has shown that physics students do not consider formulas as mathematical entities that describe phenomena dynamic nature. Therefore, the traditional aim of using formulas in their current representations was not considered. The main algebraic structures in mathematics are covariational relations called functions. Thus instead of using formulas, their function representations will be applied, because limits are traditionally used to evaluate functions. The phrase *limiting case analysis* will be used to allow transitioning of terminology between physics and mathematics. However, it will not be emphasized, and its alternative task command *evaluation of the covariational expression by taking limit* will be employed. Such undertaking is to enhance familiarity with situations when limits can be applied. While this study was published in its initial version earlier (Sokolowski, 2019), this one offers more consistent terminology and limit interpretation.

A. Sokolowski, *Understanding Physics Using Mathematical Reasoning*,
https://doi.org/10.1007/978-3-030-80205-9_6

6.2 Research Methods

6.2.1 Research Questions, Logistics, and Participants

This undertaking can be classified as a pretest/posttest one-group experimental case study (Shadish et al., 2002). The participants of the study consisted of a group of twenty-five (N = 25) college-level algebra-based physics students. Within this group, nine (N = 9) were female students and 16 (N = 16) male students. All participants possessed a prior calculus background, and they were familiar with sketching rational functions and evaluating the limits of these types of functions for the independent variable approaching an infinitely large value. They also possessed a physics background in analyzing the motion of systems of objects from high school. The pretest was assigned within the first week of the course. It was to assess students' awareness that limits are considered to quantify the formula's outputs. The results were not disclosed to students. The instructional unit was delivered within the section of dynamics about a month later. Students took posttest after 2 weeks from participating in the instructional unit. The posttests results were used to assess their gain in handling algebraic structures (functions) and tasks to learn about the system behavior. The pretest/posttest questions were similar except that on the posttest students were given a choice to use algebraic tools to answer the questions. The following emerged as the research questions for this study:

(a) *Are physics students comfortable with converting formulas to a designated covariational representation?*
(b) *Can physics students reassign the variables in formulas to a covariational relationship?*
(c) *Can physics students interpret the limit results in the scientific contexts?*

The first question was assessed using the pretest results. The posttest results were used to determine if guiding students by converting formulas to covariate relationships and introducing limits as a tool for evaluating the covariate relationships enrich their techniques of solving problems.

6.2.2 Criteria for the Study Content Selection

Analyzing the motion of a system of two objects, with or without friction, is a shared context in high school and undergraduate physics curricula rooted in a standard formula to analyze the system's motion under a constant net force.

$$\sum \vec{F} = m\vec{a} \tag{6.1}$$

When considering the mass of the system as a constant parameter, which would describe the most common scenarios, the action described symbolically as $\bullet\ \vec{F}$

directly affects the system's acceleration, $\vec{a}$ with a proportionality constant m that is the system's total mass. If mapping the formula into a typical *input-output* functional relation, its covariational scientific underpinnings could be better highlighted when the formula is presented as $m\vec{a} = \Sigma\vec{F}$. The form $m\vec{a} = \Sigma\vec{F}$ informs that the independent parameter of the covariation is the action represented by the sum of acting forces or the net force denoted by $\Sigma\vec{F}$. The result of the action is the object's acceleration with the mass m considered the system's constant parameter. Zooming more into the algebraic entanglements the acceleration of the system can be expressed as $\vec{a}(F) = \frac{\Sigma\vec{F}}{m} = \frac{1}{m}\Sigma\vec{F}$. This representation better aligns with students'daily experiences in math courses, and thus it should be easier interpreted; the higher the force, the higher the system acceleration. The traditional school laboratory exercises on exploring such motion involve deriving expressions for the system acceleration, also known as Newton's second law's operational form:

$$\vec{a} = \frac{\Sigma\vec{F}}{m} \tag{6.2}$$

Upon being introduced to this formula, students conclude that the acceleration depends directly upon the magnitude of the net force computed using the superposition principle $\Sigma\vec{F}$ and inversely proportional to the system's total mass m. These conclusions can be visualized and confirmed by generating two separate graphs, acceleration versus the net force and acceleration versus the total system's mass. Such pedagogy, albeit commonly used in physics education, embraces two different types of algebraic covariations, direct (linear) and indirect (rational) within one inquiry, which might inadvertently diminish the nature of the phenomenon's cause and effect. The underlying notion is that when investigating the acceleration, changing simultaneously the force and the mass is not convenient. For the clarity of the inquiry, two lines of investigations seem to emerge:

(a) $\vec{a}(m) = \frac{\Sigma\vec{F}}{m}$, which investigates the effects on the system acceleration of the varying total mass assuming a constant acting force.

(b) $\vec{a}(F) = \frac{\Sigma\vec{F}}{m}$, which investigates the effects of varying acting force with aconstant total system mass.

Such analyses would be recommended before students can embrace a more complex analysis of the modified Atwood machine illustrated in Fig. 6.1 that is commonly being used to investigate both covariates (a) and (b) presented above.

While for an experienced teacher, finding the system acceleration is mainly driven by setting up equations for each mass and solving the system, the scientific underpinnings and the parameters covariate entanglements have much more to offer regarding their effects on the systems' acceleration.

The problem with using a half-Atwood machine is that the hanging object m_2 (see Fig. 6.1) contributes to the system acceleration in two different natures; as an

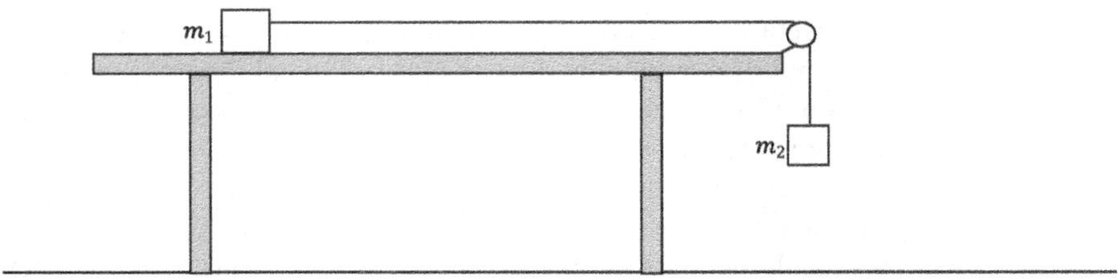

Fig. 6.1 The system of two objects connected by a massless string

inertial mass and as a gravitational mass, thus it seems that the corresponding graphs and covariates cannot be explicitly classified as direct or indirect. It is reasonable to say that its effects cannot be fully apprehended without producing a graph of acceleration when a hypothetical range of higher mass values can be considered. The effects of one of mass varying can be *observed* using the function behavior and hypothetical deductive reasoning in the form of *limiting case analyses* can be applied. Using the system to conclude that the acceleration depends directly on the magnitude of the gravitational force acting on m_2 might not precisely reflect the system's principles. While for small values of the mass, the graph appears to be linear, high range values, which enrich the analysis and extend the graph show that the graph is not linear when the force of gravity generated by m_2 is getting substantial values. Students' confusions related to interpreting experiments about Newton's laws were reported (Chief Reader Report, 2017). For instance, students faced difficulties identifying reasonable forces that affected the acceleration or could not explicitly explain their effect on the acceleration. It is seen that the bi-variational effects of the mass producing the action (gravitational and inertial) on the system acceleration can be a cause of this confusion. While this study does not aim to prove that using limits eliminates these difficulties, it is hoped that students will describe more precisely the system's motion in terms of the cause and effect when one variable parameter is considered.

6.2.3 Discussion of the Applied Algebraic Tools

While scientists and engineers commonly apply limits, using limits in school practice and examining students' ability to apply limits to enhance scientific inquiry is not often undertaken in science education research. In this study, I suggest conducting such an analysis using the context of two objects connected by a massless string (see Fig. 6.1). While the study was conducted with college students, its context is also suitable for advanced high school physics courses. In practice, such transition provides opportunities for deriving valid conclusions, especially in extreme or very small values for which direct laboratory measurements are not plausible. Although the idea seemed to illustrate applications of limits in science at first, rather than enhancing students' scientific inquiry, developed instructional units and students' responses have advanced the topic to a promising endeavor. The necessary students' math background was their familiarity with the techniques of evaluating limits of rational functions at infinity, also known in mathematics as evaluating *function end behavior*. The operational form of Newton's second law of motion when the modified Atwood machine is concerned, $\vec{a}(m) = \frac{\sum \vec{F}}{m}$ represents a rational expression, and the idea of taking limits should produce results that can be interpreted as the system acceleration. In general, using the idea of limits should extend both predictions and explanations of this type of motion that, according to Redish (2006), are fundamental goals of teaching physics. The limiting case analyzes typically do not embrace the formal processes of evaluating limits considering formulas as functions but assumes these structures (Savage & Williams, 1990). This study is posited to bring forth precise mathematics tools and apply the techniques of taking function limits in a fashion consistent with what students learn in their mathematics courses. Such an approach allows also for sketching these functions and view the limits outputs referring to these functions graphs (and asymptotes). It is hypothesized that such an approach helps students transfer their skills and understand the outputs of the processes applied in physics more clearly.

6.3 Description of the Instructional Unit

While initially, this study's intended to assess students' readiness in applying and interpreting limits in physics, the pretest results prompted to extend the mathematical part and propose an instructional unit during which the underpinnings of applying limits would be discussed in more detail. In their prior physics education, the participants studied how to construct a formula to find the acceleration using the Atwood machine model. However, it appeared that constructing it from *scratch* and guiding students on how Newton's laws can transition to more sophisticated algebraic covariation appeared as an exciting learning experience.

6.3.1 Analyzing Acceleration of the System in the Function of Mass m_2

The instructional unit was initiated by posting a problem about predicting the magnitude of the system's acceleration when either of the masses m_1 or m_2 (see Fig. 6.1) had a zero or infinite magnitude. The instructor then presented Part 1 (Sect. 6.3.1) and discussed the tasks that lead to embracing the process of concluding the answers using limits.

Problem 1 Consider a system of two blocks (Fig. 6.1). The block m_l of 2 kg is placed on a horizontal frictionless surface. The block of m_2, with an initial mass of

2 kg, is hanging over the side of the table. However, more masses are continually being added to m_2, eventually increasing its amount to an infinite value.

(a) Predict the value of the system acceleration when m_2 reaches an infinite value.
(b) Formulate an expression for the system's acceleration using Newton's 2nd law and find its value when the mass m_2 reaches an infinite amount. Use the technique of limits to derive your answer.

Students recorded their predictions, and the instructor guided them through the solution process. The inquiry was initiated by labeling primary forces acting on and within the system. The pulley was frictionless, so the system's motion was driven only by the force of gravity acting on the mass, m_2, as shown in Fig. 6.2.

By considering the forward and downward directions of the forces acting on the masses as positive and acceleration a shared parameter for all masses, the net force acting on each object was formulated.

$$\begin{cases} m_1 a = T \\ m_2 a = -T + m_2 g \end{cases} \tag{6.3}$$

Adding the equations *by sides* and factoring out the symbol of acceleration resulted in $(m_1 + m_2)a = m_2 g$. Solving the equation for the acceleration results in:

$$a = \frac{m_2 g}{m_1 + m_2} \tag{6.4}$$

Equation (6.4) represents an operational form of Newton's 2nd law. It allows for computing system acceleration by substituting known mass values for m_1 and m_2. However, converting the formula to a covariate representation and approximating the function value for a large mass m_2 will answer the problem's questions more explicitly. The pretest disclosed that the students were unsure how to handle the formula's evaluation for a considerable value of m_2; therefore, more attention to this phase of the reasoning was dedicated. It seems that classifying the parameters as dependent, constant, and independent can be conveniently processed using Eq. (6.4). Since the mass m_2 is changing, therefore this quantity will constitute the

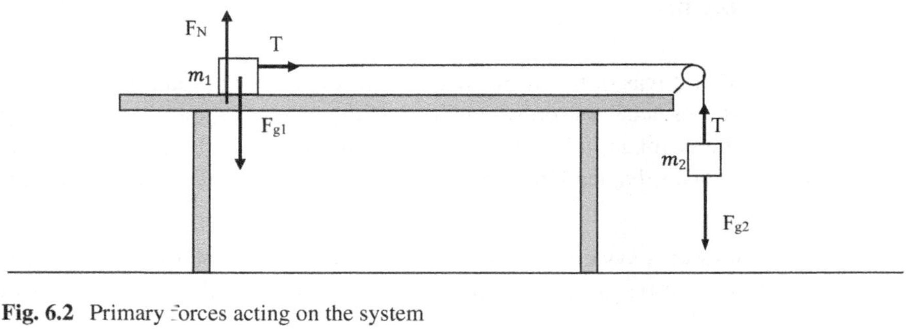

Fig. 6.2 Primary forces acting on the system

variable independent parameter of the function. To highlight the variable nature of the acceleration and dependence on the mass m_2, parameter a was embraced in a functional notation; $a(m_2)$. The mass $m_1 = 2$ kg, thus, it was replaced by the magnitude of 2. The equation then took the form of a rational covariation:

$$a(m_2) = \frac{m_2 g}{2 + m_2} \tag{6.5}$$

It is to note that the mass m_2 contributes to the system acceleration as gravitational because it is coupled with g (the numerator of the expression) and inertial as m_2 (the denominator of the expression) thus from a physics point of view, its effect on the system acceleration has a dual contribution. The effect of each depends on its relative value with m_1. Once the formula is embraced in its covariational nature, its graph (see Fig. 6.3) can be sketched, and the limits evaluated using the graph or by applying formal conduct. The limit value will be used to justify the parameter of interest; the system acceleration. Both representations should lead to the same conclusions. Referring to the graph, one will notice that when the mass m_2 increases significantly, the acceleration is getting close to 9.8 m/s². The value of the limit represents also the horizontal asymptote or the boundary of the rational function, depicted by the dashed horizontal line in Fig. 6.3. It is to note that the system's acceleration will never reach the value of 9.8 m/s² because the graph cannot cross

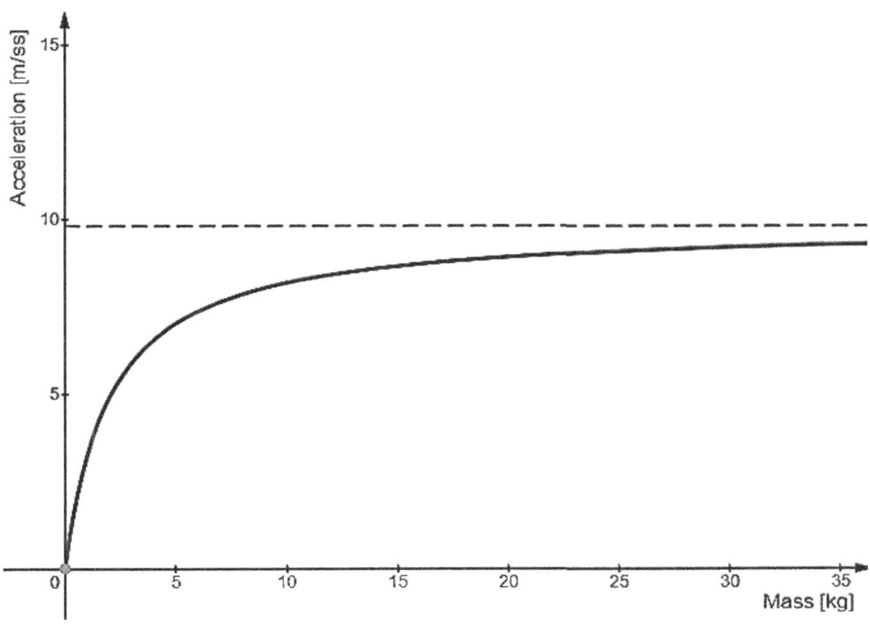

Fig. 6.3 Graph of the acceleration of the system when the hanging mass m_2 increases

the horizontal asymptote at infinity (assuming that the experiment is conducted on the Earth), and this conclusion corresponds with the system behavior.

The graph in Fig. 6.3 is generated arbitrarily for $0\,kg \le m_2 \le 35\,kg$. The value of 0 kg was chosen to highlight the physical domain of the function; $m_2 \ge 0\,kg$. While this graph appears to illustrate an exponential function due to applied restriction on its domain, it represents a rational function due to the properties of the algebraic structure. Evaluating the function for $m_2 = 0$ produces zero acceleration of the system that is represented by the vertical intercept of the graph. This situation transcends to no mass hanging, thus no system motion. It is to note that the graph provides an effective means to approximate the acceleration for various values of the hanging mass. For instance, when the mass is 5 kg, the acceleration is about 7 m/s^2. It is to note that in classroom experiments, when the values of the hanging mass are not large, perhaps when $0\,kg < m_2 < 2\,kg$, the graph appears to be linear, suggesting a direct relationship. However, this is not entirely true when the analysis extends to a substantial value of the mass. Thus, considering the mass m_2 as directly affecting the system's acceleration does not correctly reflect the system's scientific underpinnings. This is primarily attributed to the fact that at the same time, the mass contributes (increases) the total system's inertia, which transcends that the acceleration is inversely proportional to the sum of the masses m_1 and m_2. Such duality of mediating effects of the mass on the system acceleration does not emerge if the corresponding graph of a rational function is not generated.

To completely support the scientific inferences by mathematical reasoning a formal way of evaluating function limits is needed. If m_2 increases to infinity, this parameter will dominate the numerator and the denominator of the function, thus the 2 kg can be omitted as possessing an insignificant effect compared to a considerable value of m_2. The formal mathematical process is illustrated in Eq. 6.6.

$$a\left(m_2 \to \infty\right) = \lim_{m_2 \to \infty} \frac{m_2 g}{m_2 + 2} = \lim_{m_2 \to \infty} \frac{m_2 g}{m_2} = g = 9.8\, m/s^2 \tag{6.6}$$

The value of $g = 9.8$ m/s^2 represents also the equation of the horizontal asymptote of the function. The limit illustrates also a boundary value of the system acceleration, which cannot exceed 9.8 m/s^2. It is rather interesting to note that the system will never reach that value because the cart's mass cannot be equal to zero practically. In a hypothetical case, the mass can be zero. Another point is to realize that when the mass $m_2 \to \infty$, its magnitude is much larger than that of m_1, and the gravitational and inertial effects of the mass on the system acceleration are equalized and mathematically they can be canceled out from the formula $\lim_{m_2 \to \infty} \frac{m_2 g}{m_2} = g$. This means that when $m_2 \to \infty$ the system could experience a free fall. When the mass m_2 is comparable to m_1, the effect of m_2 on the acceleration of the system cannot be ignored. Moreover, algebraically such operation is not allowed either. The limiting case analysis can be employed for a large or an exceedingly small value of m_2. Indeed the sided limit evaluation leads to a similar conclusion $a\left(m_2 \to 0^+\right) = \lim_{m_2 \to 0^+} \frac{m_2 g}{m_2 + 2} = \lim_{m_2 \to \infty} \frac{0g}{2+0} = 0\frac{m}{s^2}$.

6.3.2 *Analyzing Acceleration of the System in the Function of Mass* m_1

Problem 2 Suppose that the mass m_2 remains constant (2 kg) and that the value of the mass m_1 increases to infinity.

(a) Predict the magnitude of the system acceleration
(b) Formulate an expression for the system's acceleration using Newton's 2nd law and determine its value when the mass m_1 reaches a substantial value. Use the concept of limits to support your answer.

The phases of the solution process are like in Sect. 6.3.1. Since the system's general setup does not change, its acceleration formula will also not change completely. Thus, considering m_1 a variable parameter and m_2 possessing the value of 2 kg, the formula $a = \frac{m_2 g}{m_1 + m_2}$ takes the form of $a(m_1) = \frac{2g}{m_1 + 2}$. To answer part *a* of the problem, the limit of the function for $m_1 \to \infty$ needs to be taken. Following the techniques of evaluating limits of rational functions, the mass of 2 kg can be removed as insignificant compared to a substantial value of m_1. This reasoning results in evaluating the quotient of $2g/m_1$ for a large m_1 that produces an acceleration of zero. A formal limit evaluation ilustrates Eq. 6.7.

$$a(m_1 \to \infty) = \lim_{m_1 \to \infty} \frac{2g}{m_1 + 2} = \lim_{m_1 \to \infty} \frac{2g}{m_1} = 0 \tag{6.7}$$

A similar conclusion can also be reached using graph $a(m_1)$ that is illustrated in Fig. 6.4.

Graph (Fig. 6.4) also provides a valid physical interpretation for its vertical intercept. It proves that when mass $m_1 = 0$, the system's acceleration equals in magnitude to the acceleration due to gravity 9.8 m/s^2. Likewise, Fig. 6.3 graph Fig. 6.4 can also be used to find the system acceleration for various masses of the cart and help identify and interpret its horizontal asymptote that has an equation of the limit when $m_1 \to \infty$, $a(m_1) = 0$. The asymptote is denoted with a horizontal dotted line on the graph. The horizontal asymptote can be used to conclude that the value of the acceleration will never reach the value of zero, but it will be remarkably close to zero when the mass $m_1 \to \infty$. The graph interpretations and conclusions reached using the function equation were to convince the students about its precision in depicting the phenomenon's nature. This conclusion has its scientific support in the system behavior; since the surface is frictionless, even when the mass m_1 is exceptionally large, the constant pulling force produced by the force of gravity acting on the 2 kg mass will still accelerate the mass.

Furthermore, increasing m_1 increases the system's resistance to more, decreasing the system acceleration. An actual demonstration can be used as a medium of partial experimental verification and conceptualization of this case. The graph generated in the classroom most likely will not exhibit its asymptotic behavior. Realizing this limitation, the students appreciated mathematical reasoning that extended the analysis beyond classroom experimentation.

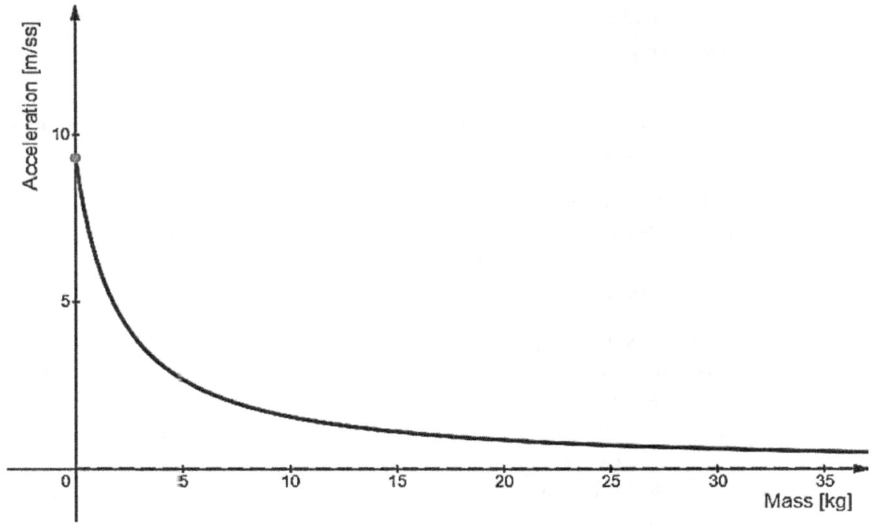

Fig. 6.4 Graph of the acceleration function if the mass m_1 varies

6.4 Data Analysis

The pretest and posttest questions were similar to the ones presented in Sects. 6.3.1 and 6.3.2. While the pretest was administered within the first week of the course, the posttests question was assigned as one of the exam questions on the cumulative chapter exam.

6.4.1 Analysis of the Pretest Results

In problem #1 about analyzing acceleration when $m_2 \rightarrow \infty$, the students derived the formula for the system acceleration; however, they could not find adequate algebraic tools to evaluate the formula when the mass m_2 increased. Even though the pretest provided clues on what mathematical processes to apply to answer the questions, the students' responses and the percent of correct answers were low. In problem 1, 20% of the students (N = 5) concluded that the acceleration would be near 9.8 m/s^2. The majority of the students, 80% (N = 19), concluded that the system's acceleration would increase, yet they did not explicate on the boundary of acceleration magnitude. Some verbatim samples of student responses are as follows: *The acceleration will increase to infinity; the acceleration will be tremendous because there is more and more mass hanging, thereby making the acceleration larger; acceleration will increase because higher mass will increase the force of gravity; acceleration will approach infinity because of the heavier weight pulled;*

acceleration is infinite; the acceleration would significantly increase; the acceleration will increase quickly. None of the students zoomed into inertial and gravitational effects that contributed to the system's acceleration. While the physics reasoning was acceptable, the students most likely recalled from their introductory physics courses that acceleration is directly proportional to the acting force; they did not attempt to explicitly transfer the formula into a covariate expression and formally take the limit. Without such association, they did not notice that the pulling mass has a dual contribution to the system's acceleration, inertial and gravitational. Some students substituted large values for m_2 and evaluated the formula, but they did not foresee the use of the idea of limits as a valid quantification tool. While all of these students did possess a background on the limit evaluation either from pre-calculus or calculus courses, the background was insufficient to realize that this problem called for applying the limiting analysis.

In problem 2, whose virtue was rooted in $m_1 \rightarrow \infty$, many students concluded that the acceleration would be zero (N = 18, 75%), which was accepted as a correct answer. However, the most accurate answer stating that the acceleration will be near the value zero but will never reach that value was concluded only by six (N = 6, 25%) students. Verbatim samples of student responses are as follows: *The acceleration will be positive but never close to zero; the acceleration will be very small because m_1 would require movement at the same acceleration; the object will have acceleration because of the gravitational field; the acceleration will be zero; the acceleration will decrease.*

In both parts of the problems, students faced difficulty supporting their reasoning using the outputs by consistently applying algebraic operations. While their physics reasoning was fine, their written work showed a lack of consistency and clarity of thought process, which illustrated their lack of algebraic tools to confirm their physics reasoning. This was explicated from their algebraic manipulations. A lack of attention to the limit interpretation could be accounted for a rare students' exposure to applying and interpret limits in their prior physics courses and perhaps also in their mathematics.

6.4.2 Analysis of the Posttest Results

On the posttest, the students were given an option to conclude the magnitude of acceleration using any method. Such formulated task was to activate their other forms of reasoning.

In problem 1, the results showed an improvement in handling the limit evaluation. The students' work showed more consistency in identifying constant and variable quantities, which guided them to be more successful. This problem was correctly answered by 54% (N = 13) of the students. Some students still faced difficulties with identifying the independent and constant parameters in the mathematical sense. Interestingly, most students who renamed the independent variable m to x computed the limit and the acceleration correctly. These students who did not

Table 6.1 Summary of Pre-test/Post-test Results

	Pretest	Post-test
Problem 1	20%	54%
Problem 2	25%	90%

rename the variables usually stated that the limit has the value of one instead of *g*. In supporting their answers, the majority (N = 20, 80%) analyzed the function's algebraic structure, and the remaining (N = 5, 20%) used simple scientific reasoning. Some verbatim samples of student responses for Problem 1 follow: *The acceleration of the system will be g because the two infinity values will cancel out; acceleration goes to infinity because increasing* m_2 *gives a high force of gravity and the system has the same inertia, if* m_2 *increases to infinity, then the acceleration would essentially just beg; the acceleration is infinity, and it will continue to increase when* m_2 *increases; the object cannot accelerate faster than g; therefore the answer g from the limit makes sense.*

In problem #2, the percent of correct answers was high, 90% (N = 22). Almost all students used the algebraic form and applied the limiting analysis to conclude and comment on their answers. This case was easier to analyze because the variable parameters were placed in the expression denominator. Verbatim samples of student responses are as follows: *As* m_1 *approaches* ∞, *the acceleration would be close to zero due to increased denominator; as* $m_1 \rightarrow \infty$, $a \rightarrow \infty$, *the larger the mass of* m_1, *the lower the rate of change of acceleration become; the acceleration would be very slow, but it would be there due to the presence of* m_2; *the acceleration of the system is 0* m/s^2 *because as in the limit, the denominator would be infinite where the numerator would become relatively insignificant, and the value would be approaching zero; inertia increases to infinity then the acceleration gets close to zero;* $\frac{2g}{\circ}$ *is going to be* 0; *as* m_1 *increases, the acceleration will be smaller and smaller.* It was noted that when asked to support their answers, the students explained how they used the techniques of limits evaluation instead of scientific reasoning. This way was not marked as a weakness against the students but it suggested including more analyses to improve on the math-physics interface. A summary of pretest/posttest results is shown in Table 6.1.

6.5 Conclusions

The posttest results indicated that the students improved their skills in applying the technique of taking limits to calculate physical quantities. The way they derived the solutions illustrated that they felt more comfortable following through the process of reclassifying parameters as constant and varying and taking limits. While on the pretest, most of these students did not realize that limits must be applied to answer these questions; upon being instructed, the students recalled the techniques and attempted to use these techniques and succeed. Their verbal supports also showed more consistency and clarity. Did such a lesson enhance their general disposition of including limits as an algebraic tool to solve physics problems? While questions targeting this objective were not developed, the students were very eager to learn how to quantify physics problems using limits. It seemed that such curriculum enhancement sparked their curiosity to explore more phenomena using that exciting math tool.

Observing students' genuine interest in knowing how limits can support physics understanding was encouraging to consider creating more contexts with such learning opportunities that are included in this book. Upon being instructed on applying limits, their solutions' processes were more consistent with what they had learned in their calculus classes. They seemed to appreciate the opportunity to integrate that knowledge because that knowledge helped them with solving physics problems. While solving physics formulas with parameters is common in physics, converting these entities to functions and learn more about system behavior is still rare that surfaced on the pretest.

Another remark voiced earlier in this chapter is that the system of two masses as set up in the problem can help learn to eliminate the misconception that the acceleration is proportional to the force of gravity acting on the hanging mass. After analyzing the acceleration graph (Fig. 6.2) and its algebraic function, claiming a direct proportionality of the acceleration was not accurate. The graph appeared linear only for limited values of the hanging mass. Still, the relation cannot be labeled as linear. Examining the graph behavior on the entire domain determined by a broader set of the hanging masses provided the evidence.

A nuance about interpreting function limits also emerged from analyzing the phenomenon. Computed function limit is typically denoted in mathematics with an equal sign (=) rather than with a sign of approaching that value (→). For example, one writes that $\lim_{m_2 \to \infty} \frac{m_2 g}{2 + m_2} = g$, not that $\lim_{m_2 \to \infty} \frac{m_2 g}{2 + m_2} \to g$ which changes the verbal interpretation. The notation of approaching that value (→) was not used during the instructional unit to assure its consistency with calculus notation. Nevertheless, an emphasis was given to limit interpretation as a value approaching the output. Indeed, if, for example, $m \to \infty$, then the output should also represent an approaching result at its end, e.g., $a(m) \to 0$. The students realized that using an arrow instead of an equal sign would more accurately describe the meaning of the limit value in the phenomenon's context. However, the notation was not augmented because the limit value in such cases shows the boundary of the quantity of interest that transcends to an equation of the function horizontal asymptotes, not to actual function value.

While this paper presented an analysis of two motion cases, the experimental setup provided opportunities for creating more variations, where the idea of inducing algebraic functions could be further explored. For example, formulating a system consisting of three different masses with two of them hanging below the top of the table would diversify the application of limits and extend the idea of modeling the properties of inertial mass. Furthermore, by applying differential calculus and taking the first derivative of the acceleration function concerning established variable mass, students could learn that the acceleration rate of change concerning the hanging mass reaches zero for a substantial magnitude of the mass placed on the table. It is prudent that formulas converted to algebraic functions have the potential to return many more exciting scientific facts about the physical system's behavior than their formula counterparts do.

References

Chief Reader Report. Accessed from https://secure. Accessed 17 Oct 2017.

Redish, E. F. (2006). Problem-solving and the use of math in physics courses. *arXiv preprint physics/0608268*.

Savage, M., & Williams, J. (1990). *Mechanics in action: Modeling and practical investigations*. Cambridge University Press.

Shadish, W. R., Cook, T. D., & Campbell, D. T. (2002). *Experimental and quasi-experimental designs for generalized causal inference/William R. Swedish, Thomas D. Cook, Donald T. Campbell*. Houghton Mifflin.

Sokolowski, A. (2019). Enhancing scientific inquiry by mathematical reasoning: Case of applying limits to model motion of a system of objects. *Journal of Physics: Conference Series, 1287*(1), 012051. IOP Publishing.

Sokolowski, A. (2020). Unpacking structural domain of mathematics to aid inquiry in physics: a pilot study. *Physics Education, 56*(1), 015009.

Chapter 7
Reconstructing Newton's Law of Universal Gravity as a Covariate Relation

7.1 Prior Research Findings

Research on teaching and learning Newton's law of universal gravitation is not extensive, perhaps because the law is mainly used to calculate the magnitudes of forces of the mutual attraction between two objects. University library search engines such as ERIC, ProQuest Educational Journals, Science Direct, and Google Scholar returned several research papers in this domain that at large related to electrostatic and electromagnetic fields and forces. Temiz and Yavuz (2014) found out that students perceive that Newton's second law did not apply in space and suggested providing more scientific evidence to change this perception. Galili (1995) stated that the core concepts of mechanics and cosmology are built strictly on the force concept and that the effects of gravitational fields are rarely mentioned. He further contended that misunderstandings of fields and forces in electricity and magnetism (E&M) stem from a "mismatch of the methodology of teaching fields in mechanics" p.380. Saeli and MacIsaac (2007) reached a similar conclusion. Sokolowski (2008) noticed that such lack of parallelism in teaching fields and gravitational forces impedes analogy as a learning method to support the introduction of fields and forces in the sections of E&M.

As exploring the concept of gravity focuses primarily on the computations of the mutual gravitational force, so do the assessment items targeting the law understanding. For example, the force concept inventory items (Hestenes et al., 1992) concentrate on testing students' ability to quantify the effects of the gravitational field on objects placed near Earth's surface. In some of the items, the gravitational field generated by the falling mass is neglected, suggesting that the only mass-producing gravitational field within this system is the larger one, the Earth, whose effect is observed. Focusing on the object on which *action-at-a-distance* is exerted without understanding the force source is also visible in electrostatics. Maloney (1985) found out that students associate the size of a moving electric charge with the

A. Sokolowski, *Understanding Physics Using Mathematical Reasoning*,
https://doi.org/10.1007/978-3-030-80205-9_7

magnitude of the force exerted on the charge ignoring the effects on the external electrostatic field. Others Cao and Brizuela (2016) and Pocovi and Finley (2002) noted that while solving problems on electromagnetic forces, students associated electric interaction as an *action-at-a-distance* due to the presence of changes rather than as an action initiated by the presence of the fields generated by the charges. Asghar and Libarkin (2010) found out that college students frequently confound gravity with magnetism and atmospheric pressure.

While students' difficulties with understanding the nature of forces associated with fields might be rooted in interpreting the phrase *action-at-a-distance,* some support the theory that the learners do not consider the forces mediated by fields, whether gravitational, electrostatic, or electromagnetic. Referring to students' difficulties in E&M, Galili (1995) contended that without proper instruction in mechanics regarding fields and forces, students would carry out their personal beliefs about the physical world in E&M. Such notion can be difficult to correct in E&M when the complexities of these phenomena do not provide much room for revisiting gravitational effects and building up the theories on these concepts. A similar conclusion was reached by Hestenes (2006). He argued that the differences underlying principles of Newtonian concepts and the students' personal beliefs are significant and that "conventional introductory physics courses are not effective at reducing the gap" p.3.

The preliminary synthesis of research points out a missing chain of realizing that an object must be placed in an external gravitational field produced by another object to experience gravitational interaction. An object produces a gravitational field, but it cannot, and it will not mediate with it. Furthermore, a lack of parallelism in teaching methodology between forces and fields in mechanics and E&M can perhaps be partially accounted for the fact that Newton's law of universal gravitation was published much earlier than when the concept of fields was introduced to physics by Faraday. Since this historical development affects how the law of universal gravitation is taught today, a brief historical perspective of how fields and forces were developed in physics follows.

7.2 Theoretical Framework

The theoretical framework is followed by the historical development of fields and forces to enrich their methodological entanglements and support the proposed didactical enterprise of introducing the law of universal gravity to students. To highlight a coherence of the methodology, it also reflects on current didactical presentations of these ideas found in physics teaching resources.

7.2.1 Historical Perspective

Gravity is traditionally perceived as the phenomenon by which objects are attracted and by which they possess weight. The methods to quantify the force of gravitational interaction between two objects were formulated by Sir Isaac Newton in 1687 and published in *The Principia* (1999). Newton's inverse-square law deduced from Kepler's laws (Hamming, 1980) supported the algebraic structure of the formula called the law of universal gravitation. Newton proved that if the orbits under a central force are conics, then the magnitude of that force varies inversely as the square of the distance. He also established the converse; if the force obeys the inverse-square law, then the orbits are conics (Hamming, 1980).

Furthermore, Newton (1999) postulated that "All bodies gravitate toward each of the planets, and at any given distance from the center of any planet the weight of anybody whatever toward that planet is proportional to the quantity of matter which the body contains" (p. 503). It is to note that the concept of fields remained silent in that theory. Without considering fields, the concept of force as *action-at-a-distance* prevailed even when Coulomb generalized the law of electrostatic interactions in the late 1700s (Smith, 2002). While discussing gravity, a need for considering fields gained its importance after Faraday and Maxwell constructed the field theory of electromagnetism in the 1800s, and it was Faraday who introduced the field concept to physics called physical *lines of force*. Maxwell agreed with Newton that the gravitational force varies as some power of the distance. "If the force were any function of the distance except a power of the distance, the ratio of the forces at any two different distances would not be a function of the ratio of the distances" (see Hamming, 1980, p.86). During most of the eighteenth century, Newton's theory of gravity remained under dispute, primarily because of the absence of a machine producing gravitational forces (Maxwell, 1873). In 1915 Einstein attempted to provide a theory for *forces- acting- at- a -distance* and described gravity not as a force but as a consequence of the curvature of space-time caused by the uneven distribution of mass (Newton, 1999). This assumption was, in fact, the critical idea of Einstein's theory of general relativity. While the theory of gravity mainly concerns the fields outside of the masses, it enables the compute the intensity of gravitational field below the surface of a uniformly dense sphere that varies linearly with the distance from its center.

The law of universal gravity derived and formulated by Newton is also presented nowadays in physics textbooks and taught in a similar way. Its algebraic structure and the theory beyond it are unquestionable as they represented the phenomena well when it was formulated. As the summary of the history of the development of fields and interactions show, new theories and discoveries after Newton implemented it to physics did not affect much the form, and the interpretation of how Newton formulated it. New developments in physics education recommending induction of covariation reasoning are not yet applied to teaching the law. This study is posited to reflect on these recommendations.

7.2.2 Contemporary Presentations of the Law of Universal Gravity

The laws of physics are built following developed models that reflect proposed theories. Newton's law of universal gravity (Eq. 7.1) is crucial because it allows to compute and predict the behavior of specific physical systems. It is seen that the conceptualization of the law, as presented in physics education resources, does not emphasize enough the effects of field interaction that Faraday has discovered. It is hypothesized that developing a solid conceptual understanding of gravitational fields will support understanding gravitational interactions and contribute to students' better understanding of electrostatic and magnetic fields.

A traditional sequence of introducing Newton's law of gravity (e.g., Giancoli, 2014) is initiated by presenting two masses, m_1 and m_2, separated by a distance r and introducing the inverse square law $F_G \propto \frac{1}{r^2}$. This statement is followed by an explanation of the inverse square law, and Newton's law is presented as Eq. (7.1) along with a conceptual description: "Every particle in the universe attracts every other particle with a force that is proportional to the product of their masses and inversely proportional to the square of the distance between the masses" (Giancoli, 2014 p.357).

$$F_G = \frac{Gm_1m_2}{d^2}. \tag{7.1}$$

Some introductory textbooks on mechanics introduce fields at the concluding phases of the course as a refinement of gravitation (see Serway, 1996 p.391). In other textbooks, the refinement points to the curvature of space rather than that of the fields. As research demonstrates, these methods do not explicitly develop such interpretation of the force -fields' interactions that would help students quantify the interaction based on the fields' magnitudes. By using direct and inverse proportional reasoning, the description of the law merely verbalizes the mutual entanglement of the parameters that constitute its algebraic structure as presented. The law supports the computation of the gravitational forces and emphasizes the masses' magnitudes as the primary sources of gravitational attraction. As such, the law formulation perhaps unintentionally hinders the effects of the gravitational fields produced by the masses on the gravitational force existence. Furthermore, since force acts on an object, the algebraic form, as informed to the learners, does not explicate what object the computed force acts and how to establish the direction of the computed force. Consequently, the law's verbal description does not explicitly help understand the theory beyond it. If the effects of fields are not discussed, the students might associate the forces with the magnitudes of the masses, claiming mistakenly that if the masses are of different magnitudes, the forces acting on each mass are also of different magnitudes (Kavanagh & Sneider, 2007). Informing students that the Eq. 7.1 computes the force of gravitational attraction does not nurture the conceptual understanding of the phenomena, which was disclosed in the pretest for this study. Following this analysis, it is hypothesized that the current presentation of the law might inadvertently imply that the primary source of the gravitational force between two objects is the existence of the masses. This study is posited that such reasoning does not reflect on the scientific underpinnings of the phenomena. Narrowing the phenomenon to its algebraic representation overshadows the effects of the gravitational fields and reduces the opportunity for linking the strength of the fields with the magnitudes of joint forces and consequently quantification of the masses' accelerations. One of the reasons for the delay in introducing these phenomena to physics can be accounted for the historical development; the interactions of fields with objects were introduced to physics almost a century after Newton formulated the law of universal gravitation. Newton did not include the effects of the fields, which might account for students' difficulties understanding the cause-effect of the phenomena. While understanding the effects of gravitational fields is not necessary to compute the forces using Eq. 7.1, it is suggested that the inquiry should move beyond these relatively straightforward algorithms. Considering the historical development, prior research findings, and pretest results, an instructional intervention (Sect. 7.5) emerged.

7.3 Methods

This enterprise can be classified as one group mixed-method study (Shadish et al., 2002). While a control group would allow for applying inferential statistical measures with effect size as the primary evaluation determinant, this was not plausible. Thus, the evaluation of the proposed method was primarily conducted by employing descriptive statistics and qualitative analysis. The instructional unit was presented to a group of 25 freshman college physics students who possessed a prior high school physics background about Newton's law of universal gravitation. Twenty of these students (N = 20, 80%) had a precalculus background, and five (N = 5, 20%) were concurrently enrolled in a calculus course. Many of these students majored in science and engineering fields. The students took a pretest during the first week of the course and a posttest after two weeks.

7.4 Didactical Underpinnings of the Instructional Unit

The purpose of designing the lesson was to emphasize that Newton's law of universal gravity is rooted in the interaction of mass m_1 and an external gravitational field produced by another mass m_2. Departing from this statement, a covariate effect of the other the mass m_2 was induced, and the formal Newton's Law of universal gravity was assembled. There was an emphasis on differentiating between the concept of gravity, gravitational forces, and gravitational fields as often these three different yet mutually related concepts, when not explicitly differentiated, might confuse interpretation. Considering the prior research and the students' pretests' level of knowledge, it was evident that a more robust conceptual underpinning and thus a modeling-type approach was needed. The modeling type approach is usually associated with an inductive inquiry initiated by specific observations that lead to reaching a general conclusion (Temiz & Yavuz, 2014). A deficiency of students' statements that interaction of the gravitational field with an external mass generates gravitational force suggested placing more emphasis on the properties of the field before introducing the traditional form of the law. To enrich the process, a reference was given to students' prior knowledge, e.g., Newton's second and third laws of motion. It was hypothesized that extending the range of tools would broaden the conceptual foundation for a deeper understanding of the law. The idea of mathematical modeling applied during the unit was to help formulate a dynamic representation of the phenomena. The following are the operational definitions used during the modeling process:

- The system consisted of one object (see Sect. 7.5.1) and then two objects (see Sect. 7.5.2) defined by their masses and spherical structures.
- The interaction among the system objects (Tipler, 2008) was mediated by gravitational fields produced by the masses and forces of gravitational attraction produced between the masses.
- The interpretation of medium of interaction; *the gravitational field* was defined as a natural phenomenon that alters properties of space around matter by giving it the ability to attract other objects placed in its proximity.

The definitions suported the development of the lecture exploratory process.

7.5 The Lecture Component

Objectives

- Develop a conceptual understanding of the gravitational field.
- Develop an understanding of gravitational force as an action generated by the gravitational field on an external mass.
- Use mathematical reasoning to understand the algebraic structure of the law of universal gravitation.

The treatment emerged as one lecture unit consisting of two parts developed in Sects. 7.5.1 and 7.5.2. In Sect. 7.5.1, a formulation of an algebraic expression to compute the magnitude of the *field intensity at a distance* produced by a spherical mass m was developed. In Sect. 7.5.2, Newton's law of universal gravity was reassembled by starting from an expression for the derived intensity of the gravitational field. The inquiry was centered on three short problems presented to the students before exploring the phenomenon underpinnings. This exploratory-type lesson allowed for highlighting to students the covariational reasoning while attending to verifying predictions. Students hypothesized the solutions/answers, and then the instructor guided the learners through an integrated reasoning process.

7.5.1 *Gravitational Field Intensity and the Effects of Covariate Quantities*

As the Earth considered a uniform solid sphere produces a gravitational field around its body, and inside it, so does any mass, even a 1-kg object. Gravitational fields produced by objects of other shapes, e.g., cubes, cylinders, are modeled using methods that include the process called in calculus the integration. The following analysis refers to the field produced outside of the sphere, thus for distances greater or equal to the sphere radius.

Problem 1 How to graphically depict and calculate the field produced by the mass?

The field must have a similar nature to the Earth's field. Thus, it must be directed toward the center of the mass, as illustrated in Fig. 7.1. The subscript "1" corresponds to the mass that produces the field. The vector on the right side was labeled $\vec{a}(d)$ since, in this location, an external mass will be placed to develop a general formula to quantify the magnitude and the direction of the force. While the symbol g denoting the intensity of the gravitational field is typically used, the symbol a was chosen to bridge the interpretation of the intensity of the field with the object's acceleration.

To formulate an algebraic representation to compute the field strength, one must determine the quantities that affect its value: the mass, m_1 of the object producing the field, and the distance, d from the center of the mass to where the field is to be computed. While the mass is the parameter describing the object producing the field, the distance is the space's parameter. The distance parameter can be considered a mediating one. It must be included in the intensity computations if different

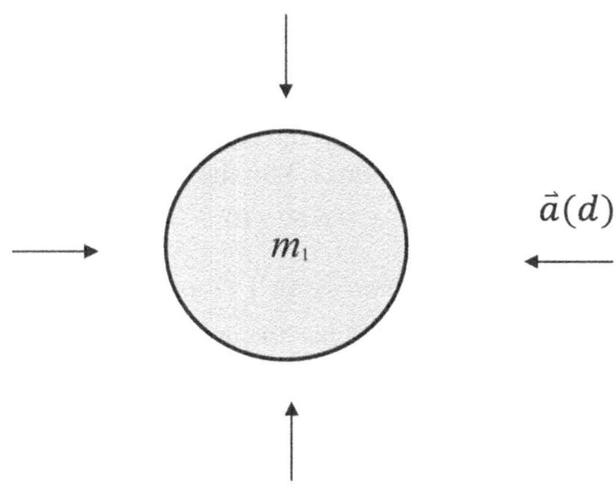

Fig. 7.1 Vectors depicting the intensity of gravitational field produced by the mass m_1

distances are concerned. In sum, direct and inverse proportional reasoning led the students' thinking toward realizing that the larger the mass m_1 is, the more intense the field is, thus using algebraic notation $a_1 \sim m_1$. Respectively the larger the distance, the lesser the intensity is. The instructor explained that the second premise was called the inverse square law and that Newton adopted it using Kepler's laws. The law can be seen as a distribution of the intensity into a spherical surface area with radius d, which supports the algebraic form of the inverse square law as $a_1 \sim \frac{1}{d^2}$. Due to a direct proportionality of the mass, the interaction between the mass and the surrounding space takes the following form:

$$a_1 \sim \frac{m_1}{d^2} \tag{7.2}$$

To convert this statement to an equation, a proportionality constant was needed. The value of the constant called the universal gravitational constant $G = 6.67 \times 10^{-11} \frac{Nm^2}{kg^2}$ was experimentally determined by Lord Henry Cavendish nearly a century later (1798) by using a torsion balance (Galili, 1995). The final form of the formula for computing the field produced by the mass m_1 is:

$$a_1 = \frac{Gm_1}{d^2} \tag{7.3}$$

Where m_1 is the mass producing the field that is the independent parameter, and d is the distance from the center of the mass to the location where the intensity is to be computed called another independent parameter that deems its value when calculating a specific value of the intensity, and G the gravitational constant that can be called a mediating parameter.

The instructor included restrictions on the function domain by stating that the function is valid only for spherical uniformly distributed masses provided the distance d be outside the radius of the sphere thus for $d \geq r$. This element was to link the formula to its functional representation.

There is an exciting conclusion reached by including the interpretation of the constant in the context of the inquiry. One can interpret it as the magnitude of the intensity of the field produced by the 1-kg object at 1 m from its center. Indeed, substituting 1 kg in Eq. 7.3 for the mass and 1 m for the distance results in the intensity of the field $a_1 = 6.67 \times 10^{-11} \frac{N}{kg}$. The intensity proves to have nearly zero magnitude. By expressing 1 N as $\frac{1kgm}{s^2}$, $a_1 = 6.67 \times 10^{-11} \frac{m}{s^2}$. The intensity of the gravitational field has the unit of acceleration m/s^2 and as such it takes a more pragmatic interpretation that will be further explored in this study. While this analysis sufficiently illustrates the field properties, expressing Eq. 7.3 as an algebraic function and graphing, it will generalize its meaning.

Problem 2 How to graphically representant the formula $a_1(d) = \frac{Gm_1}{d^2}$?

The formula consists of two independent variable parameters m_1 and d. To graphically represent it in the traditional XY coordinate system, only one independent parameter is allowed, and traditionally it is the distance d from the center of the mass, thus $a_1(d) = \frac{Gm_1}{d^2}$. A corresponding algebraic structure that can serve as a reference for sketching the graph is $f(x) = \frac{k}{x^2}$, where k is a positive constant. Considering the product Gm_1 a constant value, $a_1(d) = \frac{Gm_1}{d^2}$ resembles a rational function whose graph is a hyperbola $f(x) = \frac{k}{x^2}$ illustrated in Fig. 7.2. The graph is hypothetical; it does not include magnitudes on its axes.

Students realize that the graph is built on two different algebraic rules, linear and rational, due to possessing two different shapes; thus, a piecewise function is suitable to describe it symbolically.

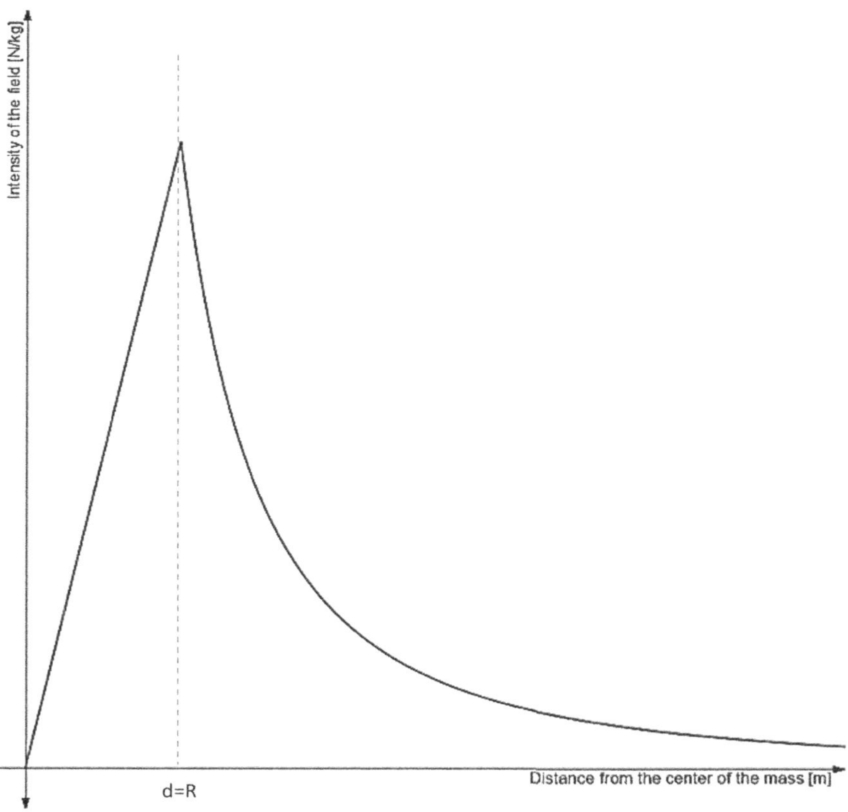

Fig. 7.2 Gravitational intensity in the function of the distance from the center for a spherical solid mass

$$a(d) = \begin{cases} \frac{Gm_1}{R^3} d, & \text{for } 0 \le d < R \\ \frac{Gm_1}{d^2}, & \text{for } d \ge R \end{cases} \tag{7.4}$$

Providing students with an expression for the intensity inside of the uniform sphere is to complete the phenomena algebraic representation. The function changes its rule at $d = R$, and this was labeled with a vertical segment on Fig. 7.2. To enrich the inquiry by mathematical reasoning, limits were used. For example, the intensity of the field at the infinite distance can be approximated by $a_1(d \to \infty) = \lim_{d \to \infty} \frac{Gm_1}{d^2} = 0$. Using a more precise interpretation of the limit value, one can claim the intensity of the field is close to zero at an infinite distance from the source producing the field. The intensity will never reach that value of zero because the expression $\frac{Gm_1}{d^2} = 0$ has no solutions when d is considered a variable parameter. This conclusion further suggests that the intensity can only be approximated to a zeroth value at an infinite distance. The analysis focused on the field interpretation when $d > R$ and a discussion of linear dependence of the field from the planet radius when $0 \le d \le R$ completed the picture. It was seen that without providing students with that part of the graph, they could mistakenly get an impression that the intensity at the sphere center has an infinite value. Such a conclusion could be inadvertently reached by the rational function behavior that is getting infinite when the denominator is getting close to zero. The scientific model to calculate the intensity inside the sphere tells otherwise; as the distance is getting close to zero, thus close to the sphere center, the field is linearly decreasing, reaching zero at the sphere's center.

7.5.2 *Reconstructing the Formula to Calculate Mutual Gravitational Force*

During this part of the discussion, the students were guided to reconstruct the law of gravitational interaction between two masses. Following a similar method, the discussion was followed by task-related problems.

Problem 3 Suppose that another mass m_2 is brought into the proximity of m_1, and it is being held there. Is there a force of gravity acting on the mass due to the field produced by m ? If yes, how to algebraically express the magnitude of the force acting on m_2 assuming an absence of any third-party force?

F_{NET} denoted the net force. However, the notation F_{12} could also be used. To add a more tangible analysis, the instructor held another 1-kg mass located 1 *m* from the existing one (in the plane of the blackboard). While this demo did not generate data, it was to illustrate its potential applicability to do so. The instructor suggested finding an expression for the net force acting on the mass at the location as indicated in Fig. 7.3.

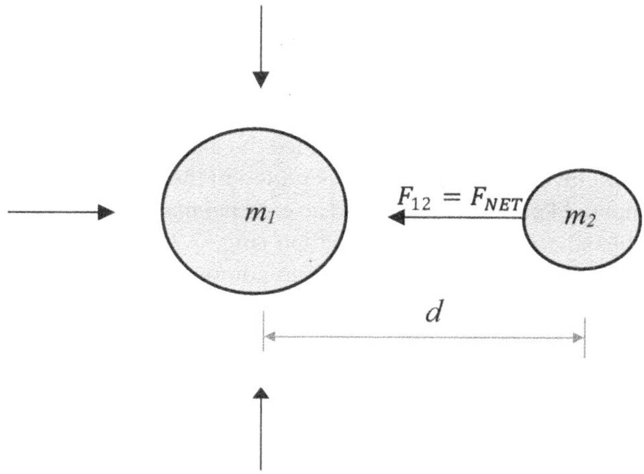

Fig. 7.3 The vector of force showing the action of the field produced by m_1 on the mass m_2

Recalling Newton's law that the net force acting on any object can be computed using $Fnet = ma$, students realize that the *force* acting on m_2, while explicitly not known at this stage of inquiry, can be found using the product of the gravitational field intensity produced by m_1 and the magnitude of the external mass m_2. The instructor emphasized that the gravitational field intensity supplied the medium for enacting the force. Nevertheless, the *strength of the field* was not the action. The gravitation field produced by m_1 coupled with the mass m_2 generated the force of action. Thus, mathematical rules paired with scientific reasoning led to a pivotal phase in the modeling process. It was hypothesized that a lack of establishing this connection earlier in the students' knowledge diverted their attention from selecting a correct response to pretest Problem 1 and 2. Omitting this reasoning essential phase might also account for students' difficulties with problem-solving when the formulation of F_{NET} acting on an object in *outer space* is considered; students do not realize the importance of the gravitational field when formulating such equations. Continuing this line of inquiry, since the intensity of the field produced by m_1 at that specific location is $a_1(d) = \frac{Gm_1}{d^2}$, and $F_{NET} = m_2 a_1(d)$, then the net force acting on $m_2; F_{NET}(d) = m_2\left(\frac{Gm_1}{d^2}\right) = \frac{Gm_1 m_2}{d^2}$. The students have emphasized the direction of the net force following the direction of the gravitational field intensity at that location. To parallel the notations with the graph, the instructor expressed the net force using a bi-numerical notation:

$$F_{12}(d) = \frac{Gm_1 m_2}{d^2} \tag{7.5}$$

It was further explained that the first digit, $_1$, of the subscript $_{12}$ denotes the *mass that produces the gravitational field,* and the second digit refers to the labeling of the external mass that is placed in the field of the first mass. While using cascade covariation and composite functions, the students realize how one parameter affects the other. Such covariational entanglement led the analysis to reconstruct Newton's law of universal gravity. Its significance is rooted in the mutual interaction between the gravitational field produced by m_1 and an external mass, m_2 inserted into the field.

To highlight the significance of the graph (Fig. 7.2), the instructor posited the following question: can the graph be used to compute force acting on any known mass at any distance from the mass m_1? The students realized that this was possible. A graph of $F_{12}(d) = \frac{Gm_1m_2}{d^2}$ could be easily generated by referring to students' mathematical skills, applying function transformation, and considering Eq. (7.3) so-called *parent function*. This part of the inquiry revealed an opportunity to develop a bi-numerical notation further and denote the force as $F_{12} = F_{NET}(d) = \frac{Gm_1m_2}{d^2}$.

To parallel the exploration, the vectors of the field produced by m_2 were labeled, and an algebraic representation for the gravitational field was formulated as $a_2(d) = \frac{Gm_2}{d^2}$. Applying the process of finding the net force on m_1; the formula took the following shape $F_{21}(d) = m_1a_2 = m_1\left(\frac{Gm_2}{d^2}\right) = \frac{Gm_1m_2}{d^2}$. This statement assured that the magnitudes of the forces of mutual attraction are equal, disregarding the sizes of the masses. This was a surprising conclusion as the pretest has shown that most students thought that the larger the mass, the larger the force is, which perhaps was accounted for the misinterpreting of the idea of *force acting at a distance*. The equity of the forces $F_{21}(d) = F_{12}(d)$, despite different masses, could also be supported by Newton's 3rd law of motion. A pictorial representation of the conclusion is illustrated in Fig. 7.4. To further visualize the differentiation between the forces and their origins, different colors were used. The red color denotes field and forces due to m_1, and the black color represents fields and forces due to m_2.

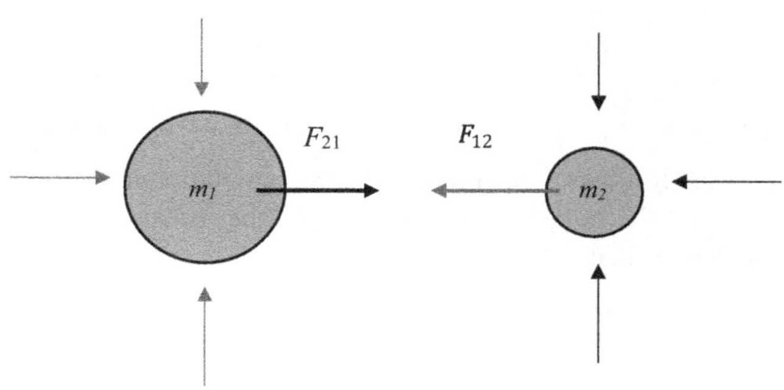

Fig. 7.4 Vectors of forces of attraction between two masses and fields

Students realized that the field is to be computed first at a distance, d, from m_2, before computing the net force acting on the external mass, m_1. While such scaffolding is not necessary for calculating forces between the objects, emphasizing this reasoning was to serve as a reference for applying the principle of superposition of fields and undertaking more advanced computations of net field and force when the system consisted of more than two masses. Extracting this operation also emphasized that the coexistence of fields' intensities plays a pivotal role in inducing gravitational force and describing objects' motion. Foremost, in the gravitational field, gravitational force can be described as an action produced by the gravitational field on another mass. It is apparent that to find the mutual force of attraction, the magnitudes of both masses and the distance between their centers are sufficient, though the modeling process was to amplify the effects of gravitational fields and to have students realize that this dominated the scientific underpinning of the phenomenon. If the mass is not present at the given location in space, that space does not produce the necessary field to attract other objects. Thus, the primary condition for gravitational force is the mass and the gravitational field it produces. Another conclusion that emerged from the analysis was that mass cannot interact with its gravitational field — only mediating between the field and an external mass placed in the field generates the force of gravitational attraction.

7.6 Analysis of Pretest - Posttest Results

While the pretest results were used to guide the instructional unit, pretests and posttest results were clustered in one section to allow for better comparisons of the results, thus allowing for a better assessment of the instructional unit.

7.6.1 Analysis of the Pretest Results

There were three pretest problems that the students were to answer. Problem #1 and problem 2 required a qualitative response, and Problem #3 was a multiple-choice question that required a selection of a correct answer. The author of this study designed these problems and consulted with colleagues who teach similar courses to target conceptual understanding of the differences between gravitational force, gravitational field, and the covariational interpretation of Newton's law of universal gravitation.

Question 1 The force of gravitational attraction between two masses m_1 and m_2 can be calculated using $F = \frac{Gm_1m_2}{d^2}$.

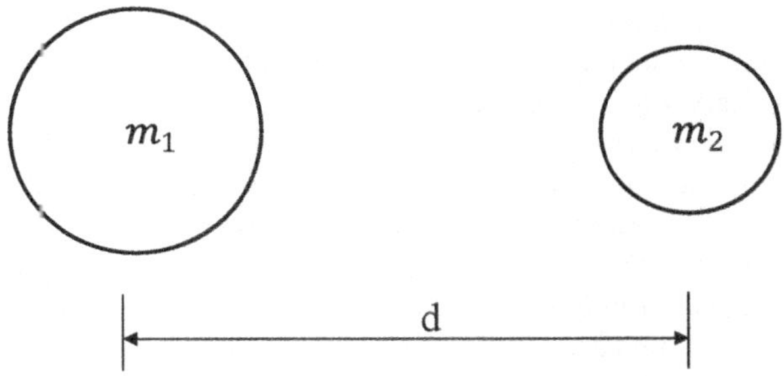

Why are the forces of the same magnitude even though one of the mass is larger than the other? Support your answer. In this diagram, $m_1 > m_2$.

The purpose of the questions was to determine if the students realize that the forces are products of two parameters mass and external fields. More specifically, the smaller mass and the larger gravitational field will produce the same force as a product of the more significant force and the smaller gravitational field.

Question 2 A mass M is held stationary, as shown in the diagram below.

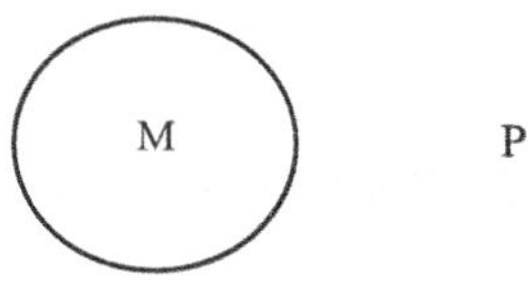

Does the mass exert a gravitational force at the location labeled P?

Students confuse gravitational force and gravitational field. The purpose of this question was to find out if there is a need to address this deficiency in the instructional unit explicitly. To avoid confusion about a presence of a mass at that location, no dot was marked.

Question 3 Two masses m_1 and m_2 are separated by a distance d and held stationary. What force is being computed by using $F = \frac{Gm_1m_2}{d^2}$?

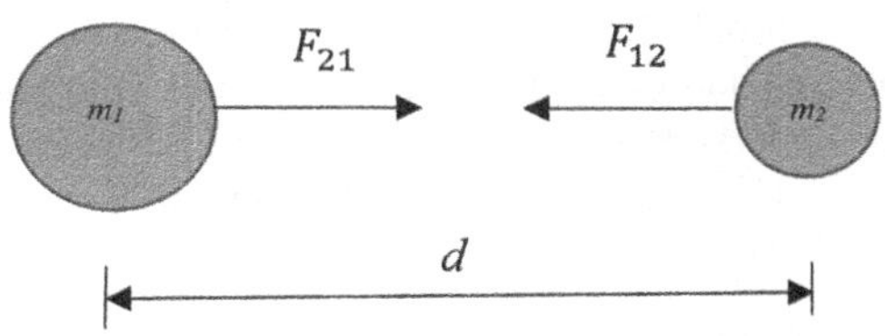

A. The net force acting on the masses.
B. Only the force that the larger mass exerts on the small mass.
C. The magnitude of the force F_{21} and F_{12}
D. None of the above because the subscripts in the general formula are not provided.

When asked for the gravitational force between two objects, students often blindly use the formula Eq. 7.1 and compute the force without realizing that they calculate the magnitudes of either F_{21} or F_{12} or each simultaneously. The purpose of this question was to find out if the students realize this nuance. The answer to this question determined the degree of details presented while deriving the equity of these forces.

Question 1 was correctly answered by three students (N = 3, 12%), and one student supported the answer by the equity of the forces. These students used Newton's 3rd law, which was accepted as correct. The underlying interaction between fields on external masses and the structure of Newton's law of universal gravity was silent in these responses. Selected students' answers are provided in Table 7.1. The correct answers were labeled with an asterisk (*).

Question 2 was answered incorectly by most students (N = 18, 72%) who suggested a gravitational force was acting at the point P, and the rest of the group (N = 7, 28%) claimed otherwise. While the question was asked to support the answer, only six students (24%) did so. Their responses are verbatim listed in Table 7.2. The first four responses (1–4) support the incorrect claim about the gravitational force's presence at that location, and the responses (5–6) support a claim of a lack of such force.

From these responses, one could infer that some of the students associate the gravitational field with gravitational force even though the question addressed the force explicitly at the location d distant from the mass-producing the field. It was inferred that the mistake could be attributed to either students' assumption that there must be a mass at the location P or due to a weak understanding of the underlying algebraic structure of the formula representing the law. None of these students attempted to support the answer by using the formula $F = \frac{Gm_1m_2}{d^2}$ and showing that it would produce a zero force when there is no other (e.g., mass $m_2 = 0$) in the proximity of m_1.

Highlighting and realizing the differences between field and force properties is fundamental for problem-solving in physics. It is concluded that the traditional textbook presentation of the law hinders the fields' effects and does not sufficiently address its fundamental effects on the force existence due to an external mass.

The correct answer for this Question 3 was choice C. Only one of the students (N = 1, 4%) answered the question correctly. Most of the students (N = 20, 80%)

Table 7.1 Students' pretest responses to Question 1

Student	Response
1	The forces are the same because the net force is zero between these objects.
2*	Due to Newton's 3rd law: action is equal to reaction.
3	The forces are the same because the value of G is constant. Each object is affected, but the force is the same. The force of gravitational pull connects these objects.
4	Because they have the same distance and gravitational pull.
5	The forces are the same because the two masses exchange forces making it so that even though one is bigger than the other.
6	There are the same forces. Just as the force of tension is the same on connected objects.

Table 7.2 Samples of students' pretest responses to question 2

Student	Response
1	Yes, there will be a small force.
2	Yes, just the downward force of gravity.
3	Yes, there is a gravitational pull on everything in space.
4	Technically, there will be a force, but very small.
5*	No, because there is no mass for the gravity to act on.
6*	No, because P does not have a mass and the gravitational force exists between two masses

claimed that the law of universal gravitation is used to compute the net gravitational force and selected the choice A. This answer was not marked as correct because net force is calculated as a sum of forces acting on a specific object, not between the objects. A small subgroup (N = 4, 16%) selected choice B. While partially correct, this choice did not include the possibility of simultaneously computing the force that the smaller mass exerted through its field on the larger one.

These results further supported the hypothesis that the problem with formulating expressions for the net force when dealing with forces generated by gravitational fields might be rooted in a lack of understanding of the cause of the force *as action-at-a-distance.* There can be other factors causing students' diversion from selecting choice C. For instance, a lack of understanding of the meaning of the net force or a lack of interpretation of the subscript notations; F_{21} and F_{12}. Nevertheless, the pretest questions' analysis suggested that the traditional method of presenting Newton's law of universal gravitation does not support students' natural intuitions about that phenomena. Hestenes (2010) argued that students' intuitions are essential cognitive resources developed through years of real-world experience and that "The chief problem in learning physics is not to replace intuitions but to *tune* the mapping to produce a veridical image of the world in the imagination" p. 27. The pretest results served as a basis for analyzing students' prior intuitions and prompts for formulating interventions. The pretest was not handed back to the students, and the correct answers were not revealed until the instructional unit was delivered and the posttest administered.

7.6.2 Analysis of the Posttest Results

The posttest questions were assigned on a summative assessment and other problems about *Rotational Motion* and the *Law of Universal Gravity.* The problems were similar in structure to assess a change in students' thinking. To reduce the effect of *fact recalling*, modifications of the question contents were made. For example, in question #2, the magnitude of the mass and the distance were provided quantitatively. In question #2, the subscripts were changed to A and B, and the symbol for the distance has been replaced by *l*.

Question 1 was correctly answered by fifteen (N = 15, 60%) of these students. While the emphasis during instructional intervention was the interaction of fields with masses, students often used Newton's III law of motion to support their answers. This law was also mentioned during the lecture but was labeled as an additional law that supports the equity of the masses. It seemed that presenting other covariate relations, e.g. if $m_1a_1 = m_2a_2$, then $\frac{m_1}{m_2} = \frac{a_2}{a_1}$ could better highlight the covariate.

Question 2 was answered correctly by (N = 18, 72%) of the students; most of them (N = 10, 56%) provided verbal supports for their answers. Out of 25 students, seven (N = 7, 28%) did not provide a correct answer. Some of the responses are verbatim listed in Table 7.3. The responses from (1 to 4 marked with an asterisks) were considered as displaying acceptable reasoning, whereas responses from 5 to 10 were marked as incorrect (Table 7.3).

Twenty-one students selected the correct answer for Question 3 (N = 21, 84%), and four students selected either choice A or C, which showed a visible improvement compared to the pretest.

A general question emerged what factors could still detour some students from differentiating between gravitational force and field (Question 2)? To find a possible explanation, these students' verbal responses were further scrutinized. Some of them (see response #6 and #10) associated the formula for gravitational field $g = \frac{Gm}{d^2}$ with the formula for gravitational force. Two inferences could be drawn from this misinterpretation; (a) students, and perhaps the instructors, often use the

Table 7.3 Posttest responses to question 2

Student	Response
1*	No, there is only a gravitational field. Until a second mass is introduced, there will be no force of gravity produced. To have a force of gravity produced, there must be two objects.
2*	There is no force acting at d = 1 *m*. There is a gravitational field from the mass, but the field needs another mass to act as a force on that object.
3*	There is no gravitational field, but there is no second object for the field to exert a force.
4*	The gravitational field is present everywhere and draws objects together, but the F_G will only be acting if there are two objects.
5	Yes, because at any distance, there is a force of gravity according to $F = \frac{Gm_1m_2}{d^2}$.
6	Yes, $g = \frac{Gm}{d^2}$ calculates the force of gravity (g) at a distance d from the mass.
7	Yes, because the gravitational field from 3 kg extends that far.
8	Yes, there is always going to be a force of gravity from that mass.
9	Yes, $\frac{Gm_1m_2}{d^2} > 0$ given *m* and d $\neq$ 0.
10	Yes, because every object has a gravitational field. The only thing is that the force of gravity at this distance so weak (used $\frac{G3}{1^2}$ to support the magnitude).

term *gravity* when they mean either intensity of the field or the magnitude of gravitational force. The lack of verbal precision might transfer to students' inability to differentiate between the force of gravity and gravitational field, (b) in the standard curriculum; students first learn the formula $F_G = mg$ to find the force of gravity, which unintentionally might suggest that when mass is used in the formula, like in $g = \frac{Gm}{d^2}$ gravitational force is also computed. A suggestion for a more profound differentiation on the mediating parameters in each equation, in $F_G = mg$; g is the external field and in $g = \frac{Gm}{d^2}$; g is produced by the mass m emerged. The effects of the mass m, are unparallel, and highlighting more the differences might help.

Several students (e.g., response #8, Table 7.3) still associated predictable action on a mass without realizing that in the problem, only one mass was given. Indeed, a qualitative study allowing for more detailed coding of students' verbal responses would allow for a more precise location of the misconception.

A high percentage of correct answers for question #3 (84%) on posttest was promising. Students realized that using the formula $F = \frac{Gm_A m_B}{l^2}$ generates the magnitudes of both forces, the gravitational force acting on the mass m_A and the force acting on m_B simultaneously. The physics curriculum does not provide many examples of such formula outputs interpretations, therefore addressing this nuance explicitly turned to be a move in the right direction.

7.7 Conclusions and Suggestions for Further Research

This study presented a blended hypothetical mathematical modeling approach of introducing Newton's law of universal gravity. Rather than the masses only, the gravitational fields coupled with the external masses were exemplified. The unit's far-reaching goal was to have students realize the differences between gravitational forces, fields, and the concept of gravity as a phenomenon. While it is true that the advantage of the field treatment might be genuinely appreciated in classical physics on account of many masses/charges, this study focused on analyzing only two masses. This was purposeful because using fewer objects reduced the cognitive load of applying advanced vectorial algebra, thus allowed focusing on the conceptual understanding. Adding more masses to the system will include a superposition principle for fields and forces but will not change the analysis's line reasoning.

The prior research highlighted students' misconceptions about gravity as one of the obstacles in teaching this physics section. It appeared that using precise language to address either gravitational field, force, or the phenomenon of gravity might, to a certain degree, support eliminating these misconceptions. The precision of indicating the quantity of interest evokes its property, units, and contribution to its understanding. the quantity described.

Linking the effects of the gravitational fields and their algebraic embodiments to calculate the force of gravity was to provide background for more advanced explorations of fields with subsequent inclusions of kinematics analysis. While the physics curriculum does not suggest more insightful analyses in these regards in introductory physics courses, students appreciated the opportunities of being introduced to methods that they will use in E&M. It was hoped that the unit equips students with methods to undertake more complex parallel computations of fields and forces in E&M.

There are certain limitations of the study. One of them is a small sample size of the experimental group and a lack of a controlled group that would allow for computing effect sizes of the treatment. It is seen that the students' posttest responses provided evidence that using this study treatment establishes the basis for further unit development and that this enterprise is worthy of considering.

This study also generated other possible research questions. For example, (a) will introducing the formula $g = \frac{Gm}{d^2}$ before $F = mg$ better highlight their scientific underpinnings and help realize the differences between the quantities each calculates and, consequently, help distinguish between the gravitational fields and forces? (b) do students mistakenly consider the product of Gm as an expression for gravitational force? (c) to what extent would the proposed introduction of Newton's law of gravity help students understand the superposition of electrostatic or electromagnetic fields?

Another line of inquiry worthy of further examining is the students' interpretation of the phrases' *forces-at-a distance* or *action-at-a-distance* that is frequently used in physics textbooks. Does the phrase *action-at-a-distance*, if not explicitly addressed and explained by the teacher, suggest to a novice learner either presence of force just due to the field without an additional mass or force produced just by mass without a field. Should such phrase be instead augmented to *forces by action originated from the presence of the field?* Such nuance, when examined, can magnify the students' misconceptions that might not necessarily occur due to their naïve perceptions but perhaps due to how these concepts are verbalized.

Although not discussed during the instructional unit, the proposed reasoning pathway also provided opportunities to differentiate between the terms *inertial* and *gravitational masses*. Can a mass, broadly defined as a property of a physical object, exhibit simultaneously inertial and gravitational nature? Indeed, there are not many real scenarios for a mass to exhibit such duality, which in short can depend on the specific nature of the mass interaction due to its position with reference to the direction of motion. If a mass causes an object's resistance to change its state of motion, it will represent its inertial properties. For example, if a net force F acts on a mass m, then the mass m represents the object's inertial mass. If the mass, m, is considered a source of the gravitational field, and thus gravitational force acting on another mass M, both these masses will experience their gravitational nature. However, the true notion of this differentiation perhaps lies in what mass is considered to be in motion. For exmaple, if the gravitational force acting on mass M is considered, M will exhibit its gravitational effects because it contributes to the magnitude of gravitational force $F = \frac{Gm_1M}{d^2}$. If the acceleration of the mass M is concerned, the M will also exhibit its inertial effects; $Ma_1 = F$. Since $Ma_1 = \frac{Gm_1M}{d^2}$, one will notice that the acceleration of the mass M is independent of its inertial mass because M is a common factor on both sides of the equation, and it can be canceled out, resulting in $a_1 = \frac{Gm_1}{d^2}$. This is instead a unique case that perhaps does not have an equivalent when the object moves along a surface.

It is seen that the differentiation between inertial and gravitational mass does not affect the understanding of the law of universal gravity but instead enhances it, and this inquiry provides another opportunity for the study extension.

References

Asghar, A., & Libarkin, J. C. (2010). Gravity, magnetism, and. *Science Educator, 19*(1), 42–55.

Cao, Y., & Brizuela, B. M. (2016). High school students' representations and understandings of electric fields. *Physical Review Physics Education Research, 12*(2), 020102.

Galili, I. (1995). Mechanics background influences students' conceptions in electromagnetism. *International Journal of Science Education, 17*(3), 371–387.

Giancoli, E. (2014). *Physics: Principles with applications* (AP ed., p. 357). Pearson.

Hamming, R. W. (1980). The unreasonable effectiveness of mathematics. *The American Mathematical Monthly, 87*(2), 81–90.

Hestenes, D., Wells, M., & Swackhamer, G. (1992). Force concept inventory. *The Physics Teacher, 30*(3), 141–158.

Hestenes, D. (2006). Notes for a modeling theory. In *Proceedings of the 2006 GIREP conference: Modeling in physics and physics education* (Vol. 31, p. 27). University of Amsterdam.

Hestenes, D. (2010). Modeling theory for math and science education. In *Modeling students' mathematical modeling competencies* (pp. 13–41). Springer.

Kavanagh, C., & Sneider, C. (2007). Learning about gravity I. Free fall: A guide for teachers and curriculum developers. *Astronomy Education Review, 5*(2), 21–52.

Maxwell, J. C. (1873). *A treatise on electricity and magnetism* (Vol. 1). Clarendon Press.

Maloney, D. P. (1985). Charged poles? *Physics Education, 20*(6), 310.

Newton, I. (1999). *The principia: Mathematical principles of natural philosophy*. Univ of California Press.

Pocovi, M. C., & Finley, F. (2002). Lines of force: Faraday's and students' views. *Science & Education, 11*(5), 459–474.

Saeli, S., & MacIsaac, D. (2007). Using gravitational analogies to introduce elementary electrical field theory concepts. *The Physics Teacher, 45*(2), 104–108.

Serway, R. A. (1996). *Physics for scientists and engineers* (4th ed., p. 391). Saunders.

Shadish, W. R., Cook, T. D., & Campbell, D. T. (2002). *Experimental and quasi-experimental designs for generalized causal inference*. Houghton Mifflin.

Smith, G. E. (2002). From the phenomenon of the ellipse to an inverse-square force: Why not? In *Reading natural philosophy: Essays in the history and philosophy of science and mathematics* (pp. 31–70).

Sokolowski, A. (2008). Using gravitational analogies to introduce electric field theory concepts—A response. *The Physics Teacher, 46*(3), 132–133.

Temiz, B. K., & Yavuz, A. (2014). Students' misconceptions about Newton's second law in outer space. *European Journal of Physics, 35*(4), 045004.

Tipler, P. A. (2008) *Physics for scientists and engineers* (5th ed., p. 367). W.H. Freeman.

Chapter 8
Parametrization of Projectile Motion

8.1 Prior Research Findings

Search engines such as Ebsco, JStar Google Scholar, or ERIC returned a substantial amount of literature documented students' misconceptions about the causes of parabolic curves of objects undergoing projectile motion. The most common misconception was that objects need some *impetus* that keeps them moving in a parabolic curve. Subsequent theories of motion provide qualitative rather than quantitative supports focusing students' attention on realizing that object's paths of motion in the gravitational field are parabolic (Jörges & López-Moliner, 2019), which might be a reason for such perception. Two-dimensional motion typically encompasses circular motion or its part (e.g., the motion of a pendulum), parabolic motion (e.g., projectile motion with varying velocity components), or an elliptical motion (e.g., planetary motion). Mihas and Gemousakakis (2007) formulated several recommendations that are to help students understand two-dimensional analysis. For instance, they highlighted the property that an object's vector of velocity is always tangential to the path of motion or that centripetal acceleration is perpendicular to tangential velocity if the tangential speed is constant. Springuel et al. (2007) also focused on students' understanding of objects' accelerations in curved paths. They applied cluster analysis to group students' answers by common traits to learn how they conceptualize vectors to depict acceleration. While objects undergoing a two-dimensional motion are typically projected from a stationary launchpad and analyzed from a stationary frame of reference, several studies conceptualized projectile motion by situating the launchpad on a moving platform (see Millar & Kragh, 1994; Klein et al., 2014). A moving launchpad was to have students realize that an object released from a moving carrier possessed an initial (usually horizontal) velocity equal to the carrier's velocity. Such situating while contextually rich added a concept of a relative motion. Many students believe (Prescott & Mitchelmore, 2005) that these projectiles do not possess forward velocity when released, and

A. Sokolowski, *Understanding Physics Using Mathematical Reasoning*,
https://doi.org/10.1007/978-3-030-80205-9_8

consequently, they think that dropped objects move backward or fall straight down. By using moving horizontal platform projectile motion displays a great deal of exploratory nature. Still, it also depends on understanding another motion; the idea of relative motion perhaps unintentionally adds another conceptual challenge to understanding projectile motion.

Several studies suggest advanced technology to have students realize parabolic shapes of the trajectories of the projected objects. For instance, Wee et al. (2012) suggested tracing the motion path by using computer software. Jimoyiannis and Komis (2001) used computer simulations and vector analysis to enhance students' conceptual understanding of the motion in the gravitational field to overcome their cognitive constraints. In a similar accord, Klein et al. (2014) used tablet computers to record and analyze a ball's motion thrown vertically from a moving skateboard. The software allowed to generate data for the object's position, plot the data to generate graphs for motion components and find their function equations. Some studies included a *damping factor* in a form of a force proportional to the square of the object's velocity with projectile motion (see Das & Roy, 2014). Espinoza (2004) sought to enhance understanding of projectile motion by linking it with an understanding of the momentum of a projected object and found out that students who studied momentum before forces were more successful in demonstrating a correct direction of the momentum, by referring to the tangential nature of the object's velocity. This idea followed Gamble (1989), who found out that reconstructing physics knowledge and making it more relevant to mechanics concepts benefits physics understanding.

Students' understanding of projectile motion is often tested on advanced nationally administrated physics exams in the USA. Reviewers of such exams voiced students' lack of clarity on understanding the different modes of movement in the horizontal and vertical direction that a projected object possesses. For instance (College Board, 2015), the reviewers pointed out that students mix kinematics horizontal and vertical components; they confuse velocities or apply acceleration due to gravity to the motion's horizontal components which reflects on their poor understanding of the motion properties.

The survey of the literature on teaching and learning projectile motion displays vast arrays of didactical enterprises ranging from eliminating misconceptions and using technology to generate mathematical representations of the motion. Nevertheless, little attention is given to have students construct algebraic representations of the motion using data and apply mathematical reasoning to uncover the motion properties. Students' abilities to describe projectile motion using mathematical reasoning appeared to be overshadowed by the conceptual description of the motion. It is seen that understanding the techniques of resolving the position velocity and acceleration functions into components is paramount to handling problem solving and graphing projectile motion component functions. It is further hypothesized that setting up the development of students' perception of two-dimensional motion using parametric equations (PE) will come across these recommendations.

8.2 Theoretical Framework

The theoretical framework is posited to take a more comprehensive view. It will begin from situating the motion in specific categories and then it will embrace it in corresponding algebraic representations. Such a view is to broaden the perspective on multidimensional motion and simultaneously generalize the patterns.

8.2.1 Categories of Motion Studied in High School and Undergraduate Physics Courses

Kinematics as a study of motion is usually initiated from one-dimensional and progresses into two-dimensional motion with projectile motion representing a two-dimensional motion. While studying circular, harmonic, or wave motion, physics resources do not necessarily inform students that these motion types can also be classified as two-dimensional and that similar mathematical tools can be used to analyze these motions. Explicating the motion type does not change the general teaching approaches, yet it would help present these phenomena as other examples of applying parametric equations and help with the generalization of the mathematical methods. Figures 8.1 and 8.2 are examples of such classifications, and they were presented to the students within the first week of study. Figure 8.1 summarizes types of motions considering the degree of freedom of object's movement and the complexity of the motion path. These categories are not exhausted, and indeed, both can be enriched. They are to help students realize a progressive complexity of the types of motion and, followed more advanced mathematical apparatus to quantify the motion.

Projectile motion belongs to a class of two-dimensional motion like parabolic motion, circular and harmonic motion. All three types of motion require applications of PE to analyze these motions' properties. One-dimensional motion called rectilinear is the most fundamental category of motion from which kinematics is usually initiated. Three-dimensional motion can include all two-dimensional motions when the third dimension is added. Another category that emerged is the possible modes of object movement (see Fig. 8.2).

When the mode of movement is concerned, projectile motion can be described as uniform in the horizontal direction and non-uniform in the vertical direction represented by parametric equations also called vector functions. Realizing these categorizations should support mathematical descriptions of the object position, velocity, and acceleration and enhance subsequent graph sketching.

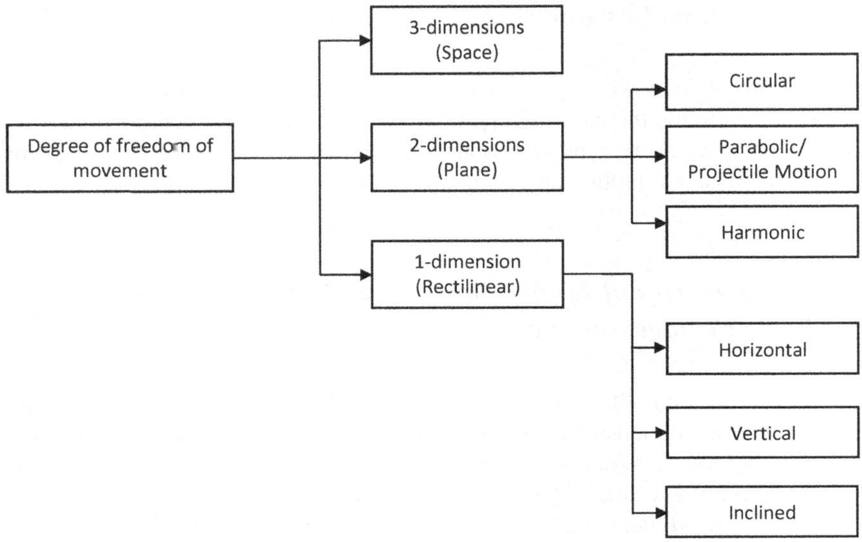

Fig. 8.1 Motion categorized by degree of movement freedom

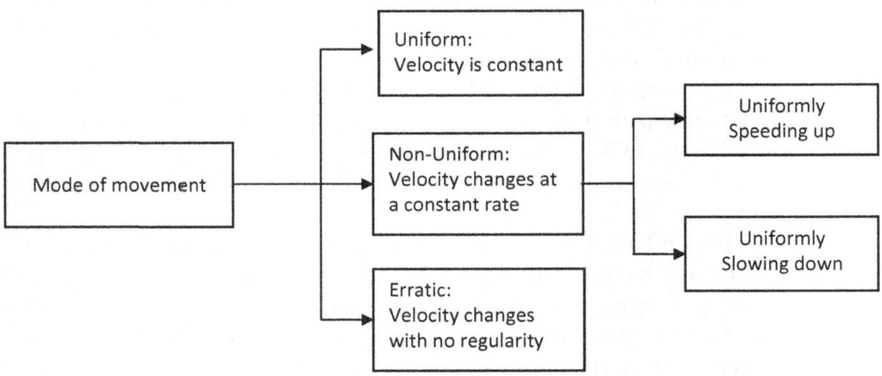

Fig. 8.2 Classifications of motion by the mode of movement

8.2.2 Why Parametric Equations?

Projectile motion is easily observable. It can be demonstrated in any classroom setting, yet the motion's scientific underpinnings, coupled with resolving it into vertical and horizontal components, are not readily observable in such demonstrations. Projectile motion is associated with motion in the gravitational field. A typical trajectory's parabolic shape might suggest that the gravitational field is the only covariate affecting the motion. It is to note that in typical problems on projectile motion, students are not explicitly directed to constructing component position or velocity functions and using these entities to answer questions. It is seen that emphasis on constructing such algebraic entities should take priority to develop in students' realities a prompt for such actions and succeed in solving projectile motion problems.

Developing this essential feature of the motion and placing more emphasis on students' skills of resolving the motion into components is not being highlighted in the current research. Therefore, to have students realize that projectile motion can be considered as being composed of two covariate modes of movement; horizontal and vertical appeared a priority. Research shows that that students mixed the components of velocities, e.g., they used horizontal components of the velocity to find the objects' vertical position. Such mistakes evidence a poor scientific understanding of the motion and perhaps a lack of foundation of familiarity with the methods of resolving the motion in components. Although taught in mathematics courses, PE properties, and their structures are not explicitly highlighted in physics textbooks and curricula, and thus, research on using PE to understanding motion is limited. Resolving the motion into components is typically rooted in finding the initial velocity components and enacting subsequent formulas. Mathematizing is reduced to formulating equations without deeper elaborations while it is possible to do that. While the component position functions $x(t)$ and $y(t)$ present an excellent opportunity to use covariational reasoning to support the phenomena' interpretation, this opportunity is lost.

This study emphasizes constructing a mathematical representation of projectile motion considering the motion's parametrization as a determinant at hand. Its main flow is designed using the scheme developed in Chap. 4. However, the learning theory chosen to guide the treatment design and the students' knowledge acquisition is the *constructivist learning theory,* whose foundations are summarized in Sect. 8.2.3.

8.2.3 Foundations of Constructivist Learning Theory

The cognitive effects of constructivism are highly appraised and advocated by von Glasserfield (1991). Constructivism is a theory in education that prioritizes gaining knowledge by constructing new understandings and integrating it with what learners already know (Jonassen, 1991). Knowledge acquired by constructing is more accessible, absorbed, stored, and retrieved from learners' long-term memory. Following this theory, Clark and Mayer (2016) suggested that effective knowledge acquisition is characterized by active processing. Learning occurs when people engage in appropriate cognitive tasks that require organizing provided information into coherent structures. While this learning theory is not visible in physics education, it is often applied in mathematics, helping students make sense of abstract mathematical structures. Since the structure of PE will be brought up during the treatment, using this theory to help students understand the suitability of these algebraic structures emerged as a prompt in the right direction. The students will actively construct PE for a given path of a projected object generated by a simulated experiment. Thus, students will derive algebraic equations for the motion using data generated from the observable experiment. Such task structuring creates appealing conditions for merging scientific and mathematical reasoning, thus nurturing new knowledge.

The advantage of the simulated dynamic representations is their potential also highlight the time parameter, *t*, and the combinations of *x* and *y* components of the vectorial kinematics quantities to form the object's path of motion. Observing how a change of time affects the object's position in both the *x* and *y* directions should highlight the mutual effect of time on both these motion components.

While learning or constructing mathematical structures involves manipulating mathematical symbols, graphical representations, especially their dynamic embodiments, have the highest potential to help students realize their scientific meanings. Being exposed to graphical representations, the learners acquire a set of tools that expand their capacity to model and interpret physical, social, and mathematical phenomena (NRC, 2012). Formulating internal or mental representations through understanding their external embodiments and retrieving their mental pictures plays an essential role in communicating messages in science and mathematics. Von Glasersfeld (1991) posited that the environment is construed by how one's internal representations are formulated. Enabling these experiences by engaging and intellectually stimulating learners through carefully designed learning environments deems nurturing effective learning.

8.3 Methods

8.3.1 Study Description and the Research Question

This study can be classified as one group experimental with posttest evaluation (Shadish et al., 2002). While a pretest would allow for a more accurate treatment evaluation, some of the study participants did not possess a physics background. Thus, the pretest would not objectively reflect on their prior physics experiences and provide reliable data for the treatment evaluation. The following question guided the study:

- *Will situating the learning of projectile motion as a two-dimensional motion representation help students realize its mathematical and scientific underpinning?*

The evaluation of the research question was assessed by analysing students' responses to the following qualitative questions:

- ***Question 1:*** *Projectile motion can be analyzed using its components. Evidence of what experiment/demonstration from the listed below convinced you that the motion has different modes of movement in the vertical and horizontal direction:*

  (a) *Observing the object move along its trajectory.*

(b) *Working on the virtual lab and using the path of motion to resolve the path into component function.*
(c) *Other experience from an introductory physics course.*

- ***Question 2:*** *Explain the process of analyzing mathematically projectile motion.* It was concluded from the prior research findings that confusion with using *x* and *y* components of the motion resulted from a poor understanding or a complete lack of possessing skills of constructing and using parametric functions to represent projectile motion. Therefore, mathematical reasoning and property of parametric equations came forth in this study. Students answered both questions after completing the treatment.

8.3.2 The Participants

Twenty (N = 22) first-year university physics students participated in this study. Seventeen of these students (N = 17) had a prior precalculus background and studied PE in their math classes, ten of these students (N = 10) were enrolled concurrently in a calculus course. Twelve of them (N = 12) were males, and eight (N = 8) females. Most of these students possessed a prior high school or college physics background.

8.3.3 Lecture Component Sequencing

Concepts in mathematics are usually introduced as context-free; thus, presenting these ideas as supporting physics understanding is always welcomed by physics students. While these students did have a precalculus background where these ideas are taught, PE and their applications were introduced during the first days of studying kinematics in this physics course. Upon introducing PE, these structures were used periodically to enhance one-dimensional motion by including the missing motion dimension; thus, $x(t)$ or $y(t)$ even though they were settled as either $x(t) = 0$ or $y(t) = 0$. Introducing the concept of parametric equations before studying projectile motion was to reduce its cognitive load and to better prepare the students to apply these mathematical structures in more advanced scientific contexts. The idea was introduced gradually, and its phases were embedded as they fitted the traditional scope and sequence of kinematics. The learning objectives embedded to enrich the traditional kinematics curriculum by using parametric equations were the following:

1. Enriching description of one-dimensional motion by using two-dimensional representation.
2. Resolving into components a uniform motion of a car moving in two-dimensions.

3. Constructing position functions for projectile motion given by its path of motion.
4. Constructing position functions for various types of two-dimensional motion.

8.3.4 Topics Embedded within the Curriculum to Enhance the Treatment

Objectives (1), (2), and (3) were considered prerequisites to Objective (4), which constituted the core lab activity of the study. It was hypothesized that inserting simple two-dimensional motion cases within the existing curriculum will provide students with the necessary mathematical tools to understand the algebraic underpinnings of the projectile and, eventually, circular, and harmonic motion. The subsections that follow describe how each of the objectives was introduced.

Objective 1 Enriching description of one-dimensional motion by using a two-dimensional representation.

PE is one of the objectives covered in a traditional Precalculus course. Thus, a reference to these concepts can be made. Could this be considered a prerequisite to apply these math tools in physics? Not necessarily. While vector components can depict instantaneous position, velocity, or acceleration, PE in physics courses can be considered a dynamic extension of vector components representation that allows for computing instantaneous values of these vectors at any time instant during the object motion, thus they can be called parametric functions (PF). The dynamic representations were further mapped into algebraic functions, with time t, considered the independent variable. The terms *component functions* and *parametric equations* were used interchangeably. PF cannot be narrowed to motion analysis only; in general, they can be considered mathematical tools that help describe a phenomenon from different perspectives. For example, a temperature change of water in a beaker can be measured simultaneously by two different thermometers, one in Celsius one in Fahrenheit scale, leading to concluding temperature conversion equations. Object's position undergoing two-dimensional motion can also be viewed using its component position functions, typically in the horizontal and vertical directions. In short, once the correlation between math and physics was established, the terms *parametric equations* were not used but instead, the instructor called these entities functions depicting components of an object's path of motion or components representing velocities or accelerations in both directions x and y.

More precise lines of thoughts on how PF were introduced to the physics curriculum are presented in the examples that follow. In short the learners were first presented with scenarios of various paths of object motion. A discussion of how to algebraically describe the motion led the students' thinking toward realizing that a need for more precision in motion descriptions is compulsory. While such motion descriptions are not being practiced, these didactical enhancements were purposeful on the premise that understanding the main idea of using PF will be more accessible when applied to simple motion cases. After such motion descriptions were understood, students were introduced to cases where position functions, in x and y directions, took non-zeroth forms. In any of these examples, it was essential to highlight the direction of the view used to formulate respective functions. These elements were considered the basis for enabling students to envision components of the motion and construct internal mental images of the functions and formulate symbolic representations. The view is traditionally established from the sides or /and from the top. Respective arrows labeling corresponds to the axis on which the object's position is projected. In all cases, the students were also supposed to classify the motion according to provided categorizations using Figs. 8.3 and 8.4.

Example 1 Starting from 4 *m* left, a student walks for 2 minutes along a horizontal line with a constant velocity of 2 m/s in the right direction. Formulate functions to describe the student's position in the horizontal and vertical directions.

Establishing a reference frame is necessary because that is how students initiate formulating functions in mathematics (Sokolowski & Capraro, 2013). The direction of view (projection) was also indicated to have students realize that this element helps build the mental model of the anticipated type of motion and, consequently, the function's algebraic form.

The position functions of interest were $x(t)$ and $y(t)$. The labeling in Fig. 8.3 was used to find $x(t)$ and Fig. 8.4 supported the finding of $y(t)$.

How to formulate $x(t)$? A constant rate of change of these parameters expressed as the student's speed implies a linear function $x(t) - x_i = v(t - t_i)$, where (t_i and x_i) are the coordinates of the initial point of the function. Substituting the given coordinates, the function takes the following form $x(t) = 2t - 4$. How to find $y(t)$? First students needed to formulate a mental view of the student's movement along the y-direction. Thus, determine if it changes and if it does, how it changes? It is apparent that the student's position as viewed from the left side of the picture does not change and can be described as $y = 0$.

To assure a consistent notation, the vertical position of the student was denoted as $y(t) = 0$.

In sum, both functions precisely describe the student's position while he is walking. While in math classes, the domain for PE is required to fully parametrize the motion, the time of motion $0 \leq t \leq 120s$ was also included. This helped realize that components of the motion share the same domain or that the time interval for both motion components is the same. This element will be more significant when the projectile motion will be discussed. A complete description of the student's position as he walks is:

$$\begin{aligned} x(t) &= 2t - 4 \\ y(t) &= 0 \qquad 0 \leq t \leq 120s \end{aligned} \tag{8.1}$$

An element visible in the diagram but not typically included in physics textbooks is the XY coordinates. The Cartesian plane does not often accompany physics

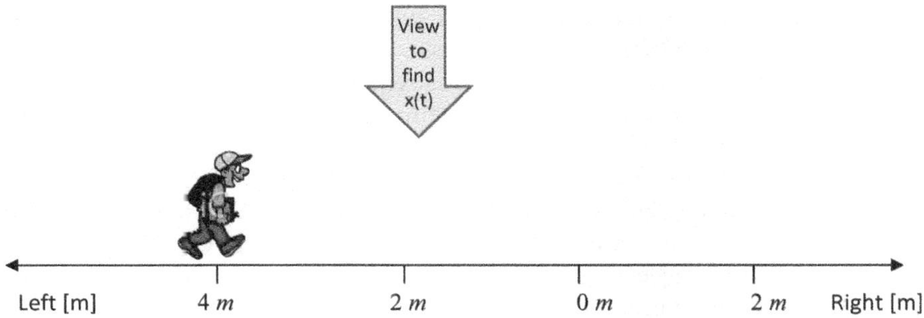

Fig. 8.3 One-dimensional frame of reference to determine the student's horizontal position

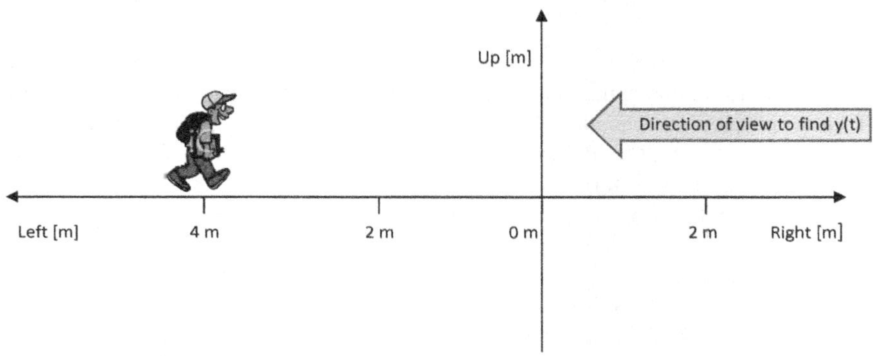

Fig. 8.4 Two-dimensional frame of reference to determine the student's position

formulas because physics formulas as static representations do not require such graphical representation. To describe projectile motion, the inclusion of the axes of projection will help construct a mental picture of the motion mode and conclude the type of algebraic functions to mathematize them.

Example 2 A baseball is projected from the ground with an initial velocity of 20 m/s. Referring to the coordinates provided, write PE for the object's position.

This experiment was demonstrated to students using the simulation (Fig. 8.5). It displayed a *free fall* vertical motion with horizontal position unchanged yet not positioned at $x = 0m$ but at $x = 2m$. Using a formula for uniformly accelerated motion in the vertical direction and a constant function in the horizontal, the students learned that:

$$\begin{aligned} y(t) &= 10t - \frac{9.8t^2}{2} \\ x(t) &= 2 \qquad\qquad 0 \le t \le 2.0s \end{aligned} \tag{8.2}$$

These functions mathematized the position of the projected object. To complete the representation the total time of the motion was calculated and included in the math description. Since some of the students were concurrently taken a calculus class, this example provided an excellent opportunity to apply their knowledge. Taking the derivative of each function, they derived the velocity functions:

$$\begin{aligned} v_y(t) &= 10 - 9.8t \\ v_x(t) &= 0 \qquad 0 \le t \le 2.0s \end{aligned} \tag{8.3}$$

Differentiating the equations again resulted in a set of acceleration functions:

$$\begin{aligned} a_y(t) &= -9.8 \\ a_x(t) &= 0 \qquad 0 \le t \le 2.0s \end{aligned} \tag{8.4}$$

These examples supplemented traditional textbook questions when one-dimensional motion was concerned. While finding both component functions was not necessarily, these examples were to expand students' view on motion and train them in considering two-dimensional analysis as richer algebraic representations to depict objects' position, velocity, or acceleration.

Objective 2 Resolving into components a uniform motion of a car moving in two-dimensions.

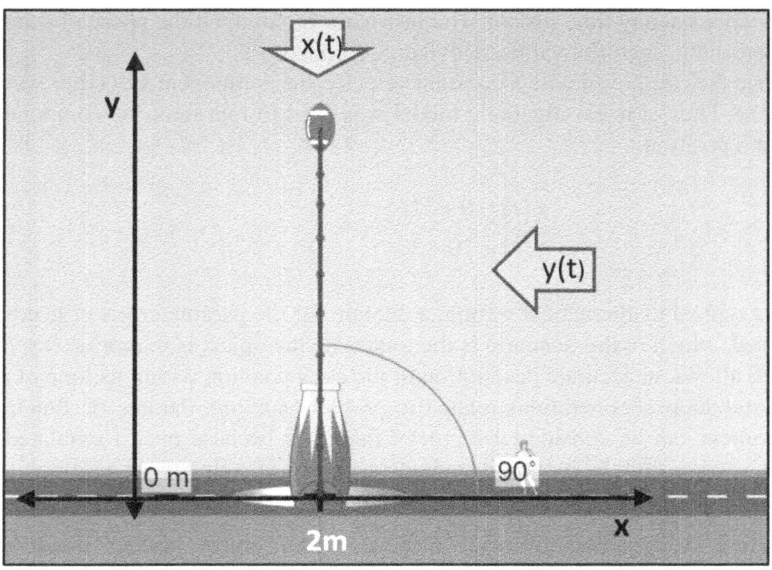

Fig. 8.5 Parametrization of vertical motion. (Source: PheET Interactive Simulations (n.d.))

The following example was discussed with students in the concluding lesson on one-dimensional motion. Its purpose was to present another technique of analyzing two-dimensional motion that involved an object's motion with a constant object yet at a certain angle concerning the horizontal axis. This case provided an excellent opportunity for extending vectors' meaning and linking these representations with their corresponding parametric functions.

Example 3 Starting from 20 km, North, a car moves in the direction 30° North of East at 60 km/h for three hours. Construct functions for the car's position along the East and North directions.

The bird's eye view of the motion is provided in Fig. 8.6. The axes of projection will be sketched *on the ground*; however, they will retain their orthogonal character.

The Cartesian plane's labeling is represented by a geographical reference in which the motion path was embraced. In this example, finding components of all vectorial quantities involved in describing the motion was necessary. The vectorial quantities in this scenario are the car's velocity and position. Thus, one needed to find the car's velocity components in the East and North directions and the components of the car's initial position in these directions.

The velocity components were: in the east direction $v_x = (60km)\cos 30^o = 52\frac{km}{h}$ and in the north direction $v_y = (60km/h)\sin 30^o = 30\frac{km}{h}$. The components of the initial position were: in the east direction $x_i = 0$ and $y_i = 20\ km$ in the north direction. The components of position and velocity were used to construct respective component functions. The vector nature of these quantities was changed to their dynamic functional representation that would, in return, allow for computing their instantaneous values at any time instant. The instructor highlighted the phase of transitioning from static vectorial values to dynamic forms of PF.

Since the car moved with a constant velocity, the component velocities were also constant. Thus, a linear algebraic model was used to represent the components of the car's position:

$$\begin{aligned} x(t) &= 0 + 52t \\ y(t) &= 20 + 30t \qquad 0 \le t \le 2h \end{aligned} \tag{8.5}$$

In a typical mathematical setting, a domain for the parameter t is or needs to be provided, which in this scenario is the total time the object is in motion $0 \le t \le 2h$. The PF allows an accurate description of the car's position within its time of movement and supports operations related to problem-solving. Parametric functions in this context can be considered *vectorial functions* because once formulated; they can be used to find the motion instantaneous position, velocity, or acceleration that possesses due to definite direction vectorial natures.

Objective 3 Parametrization of motion of an object sliding down on an inclined plane.

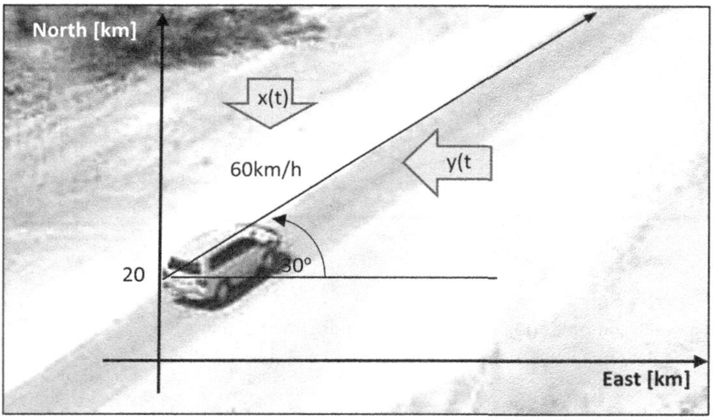

Fig. 8.6 Parametrization of inclined motion

Example 4 An Object is sliding down on a frictionless incline plane of height h and an angle of inclination θ. Establish a frame of references and write parametric equations for its position, velocity, and acceleration when it moves along the inclined plane.

This example provided students with an experience of finding components of the motion when the object's velocity was not explicitly provided and merged three representations: experimental, diagrammatic, and algebraic. This case was demonstrated to students by the instructor, and the students were guided to parametrize the motion. The available measuring devices were a ruler and a stopwatch. The instructor explained the lab's purpose and initiated the discussion on the tasks that must be taken to construct the motion position functions in x and y directions. Typically, the x axis is positioned along the incline, however standard positions of X and Y coordinates were selected to provide a bridge to the parametrization of projectile motion. Figure 8.7 depicts a diagram of the scenario's side view; the corresponding views and projections' axes are also labeled.

This example reinforced the idea of formulating all: position, velocity, and acceleration functions in x and y directions. Thus, vector components of all, acceleration, velocity, and initial position, were to be found beforehand. The process of parametrization consisted of two stages, first using general formulas for acceleration and constructing two functions that depicted the motion and then finding necessary numerical values. The students observed the motion and realized that the block accelerates uniformly along the incline. They realized that the block's acceleration could be computed by knowing the distance (the length of the ramp) and motion time. Computed acceleration was positioned along the incline. Therefore, it needed to be resolved along the x and y axes. Similarly, the velocity and the initial position needed to be embraced. The object accelerated uniformly thus, functions for uniform accelerated motion needed to be employed.

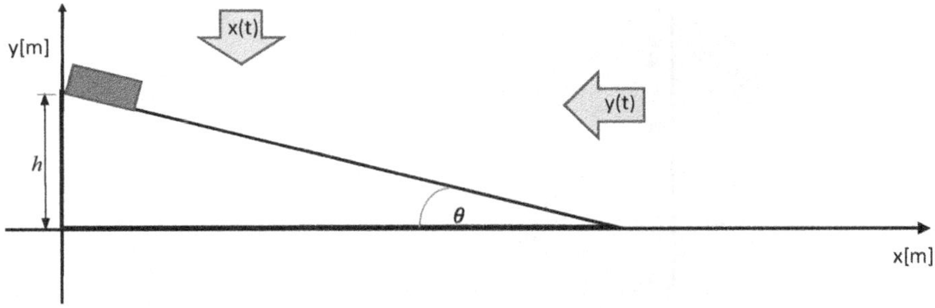

Fig. 8.7 Parametrization of object's motion on an inclined plane

The object started from rest, thus $v_x = 0$ and $v_y = 0$. By finding the angle of inclination needed to find the components of the acceleration a set of parametric position functions emerged:

$$x(t) = \frac{a(cos\theta)t^2}{2}$$
$$y(t) = h - \frac{a(sin\theta)t^2}{2} \qquad 0 \le t \le t_f \tag{8.6}$$

Students then developed velocity and acceleration component functions. Calculus students taking this physics course were asked to find the velocity and acceleration functions using the derived position functions. The parameter t_f represented the total time of the block's motion on the inclined plane. When the functions were formulated, an exciting discussion emerged about the effect of the angle of inclination on the shapes of the PF and their graphs. The angle affected the acceleration's value in both directions; thus, it also affected each parabola's vertical stretch or compression when the computed functions were considered the parent functions.

Objective 4 Using PF to construct functions for projectile motion given by its path of motion. This objective constituted a separate lab activity that is described in Sect. 8.4.

8.4 General Lab Description

The students conducted this activity in the physics lab. They took data by being provided with a snapshot of a simulated motion of a projected object demonstrated to them. They worked in groups of four to enable dialogs.

8.4.1 Lab Logistics

The idea of learning the properties of projectile motion considering its components is reminiscent of Galileo's approach to solving projectile motion problems.

However, by constructing covariate relations and realizing these functions' algebraic properties, students were to discover the unique properties of the motion when components are concerned. Linear position function represents motion with a constant velocity, and a quadratic position function represents motion with a constant acceleration. Attributes of functions such as x – and y-intercepts, slope, or maximum value coordinates were also considered. The continuity of reinforcing these properties was to establish a general strategy to mathematize multidimensional motion. The emphasis was placed on discovering the algebraic functions' attributes and using these attributes to determine the motion properties. According to the proposed modeling scheme (Part 2, Chap. 4), the general stem of instructional support was formulated focusing on merging algebraic structures and their scientific interpretations in the context of projectile motion. In its general form, the activity could also be suitable for other college-level physics courses. The differentiation would stem from the complexity of the motion under investigation. While projectile motion is often considered a phenomenon that helps students understand motion in two dimensions (see e.g., Jordan et al. 2018), in this activity, students used the techniques of deriving position functions for a two-dimensional motion to learn about specific properties of the projectile motion. Thus, a projectile motion was considered one of the multidimensional motion types rather than one supporting an understanding of two-dimensional motion.

8.4.2 Gathering Data to Construct Positions Functions for a Projected Object

The snapshot of the simulation (Fig. 8.8) was displayed in the physics lab using the classroom screen. Displaying the simulation on a classroom screen provided opportunities for directing students' attention to specific motion attributes and enhancing discussions. The simulated lab was used to generate a path of motion of an object projected at different angles and realize its different shapes. The selected path of motion was used to derive the object position in both the vertical and horizontal directions.

To link this motion with the types of multidimensional trajectories, the instructor posited a question about the motion classification considering the degree and mode of movement. All students considered this motion as two-dimensional, however realizing the mode of movement in each direction was not that apparent for all students. The instructor then directed the discussion towards finding out how to algebraically formulate the projectile motion's position functions in the horizontal and vertical directions. Some students were inclined to claim that the object accelerates in the vertical direction due to gravitational field and does not in the horizontal; however, supporting these claims by mathematical reasoning and algebra tools

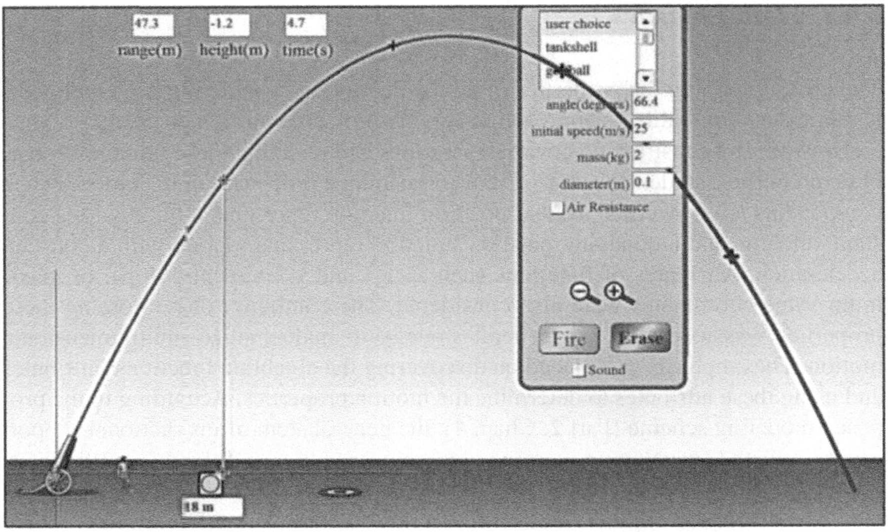

Fig. 8.8 Technique of data taking to construct parametric functions for projectile motion. (Source: PhET. Interactive Simulations (n.d.))

was not that apparent. Some considered finding components of the initial velocities but could not foresee a need to check the values of these velocities over the entire span of the motion. The question remained open about proving that the object does not accelerate in the horizontal direction. The discussion concluded with generating the table of values for the positions in x and y directions, finding these functions algebraic forms, and using these functions' attributes to determine movements' mode in each direction.

How to generate the table of values for these functions not being provided with these functions algebraic representations? To conclude the method, the learners were directed to positions of the projected object marked by (+) along the trajectory (see Fig. 8.8). A helpful element was the embedded timer that marketed the objects' positions after each second of the motion. Students realized that this attribute could depict the functions as a table of coordinates and then find their algebraic equations. The directions of view, while not labeled, remained the same as in the previous motion diagrams. Thus the top and side view were used. Figure 8.9 provides more details on how the snapshot was used to generate the data.

The following sequence of tasks was provided to students to generate the data:

(a) On the parabolic trajectory, label the position of the object after each second of motion as A, B, C, and D.
(b) Measure the positions OA_x and OA_y and formulate tables of positions in each direction separately for the remaining highlighted points.

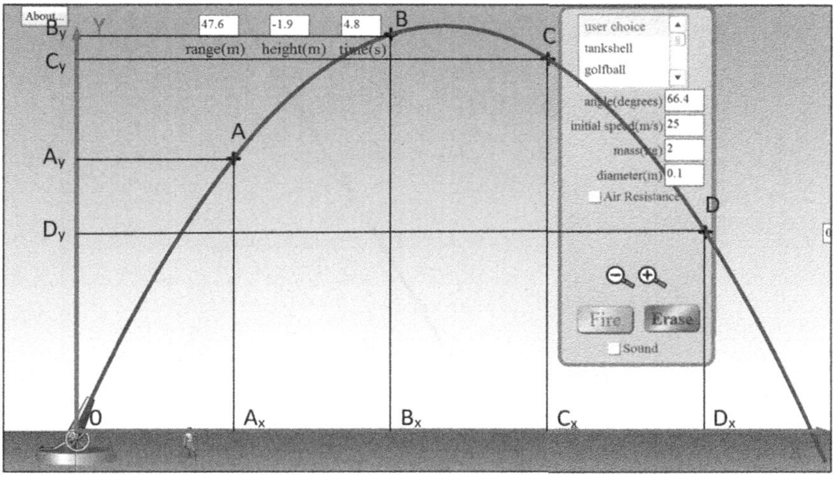

Fig. 8.9 Measuring the vertical and horizontal position of the object after each second of motion. (Source: PheET Interactive Simulations (n.d.))

Table 8.1 X and Y coordinates of the projected object

Time (s)	0	1	2	3	4	4.8
Coordinate	(0,0)	A	B	C	D	E
Position X(m)	0	10.3	21.1	30.1	40.4	48.5
Position Y(m)	0	18.6	26.1	24.5	13.3	0

Students realized that as on the *x*-axis, the position progresses with a constant rate of increase; on the vertical axis, the position increases and then decreases (Table 8.1).

The students also realized that the time variable linked the horizontal and vertical positions; they noticed that both motion components shared this variable. Following the data's progression, they further confirmed that the horizontal position changes by a constant rate, whereas the vertical position increases and then decreases, indicating an acceleration. These predictions were verified after the algebraic representations of these data tables were found.

8.4.3 Constructing Representations of the Position Functions

Functions' visual representations are the most convincing for the students (Gates, 2018). Thus, the next step of the inquiry was to graph $x(t)$ and $y(t)$ independently, realize the functions' attributes and thus motion properties in the designated directions. Students were prompted to draw the best-fit curves for each data set and find the function equations by the methods studied in their mathematic courses.

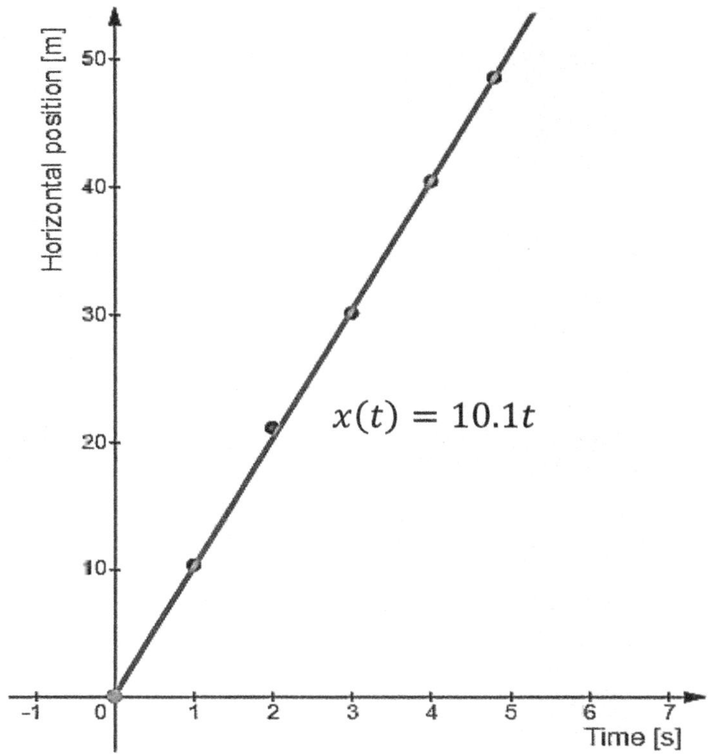

Fig. 8.10 The best-fit line and its equation for the horizontal position of the object

Figure 8.10 illustrates the graph of the object's horizontal position versus time, and Fig. 8.11 shows the graph of the vertical position of the object versus time.

The graph (Fig. 8.10) provided ample evidence that due to a constant slope of the horizontal position function, the rate at which the horizontal component of the object position changes is constant. Furthermore, the best-fit line returned the slope of 10 that represented the horizontal component of the velocity $v = 10\frac{m}{s}$. Thus, $x(t) = 10t$ showed the position function in the horizontal direction.

The best-fit curve uncovered a quadratic form for the vertical position, proving its accelerated mode in that direction (Fig. 8.11).

The students realized that this graph does not represent motion with a constant velocity because it did not resemble a linear model. The projected object was under the influence of the constant gravitational field, decreasing the object speed at a constant rate of 9.8 m/s^2 in the upward direction and increasing the speed at a constant rate of 9.8 m/s^2 in the downward direction. The students realized that the coordinates of both functions $x(t)$ and $y(t)$ generated coordinates for the motion path shown in Fig. 8.9. The complete algebraic representation of the motion is given in Eq. 8.7. It does include the time interval for the motion to follow up on the general

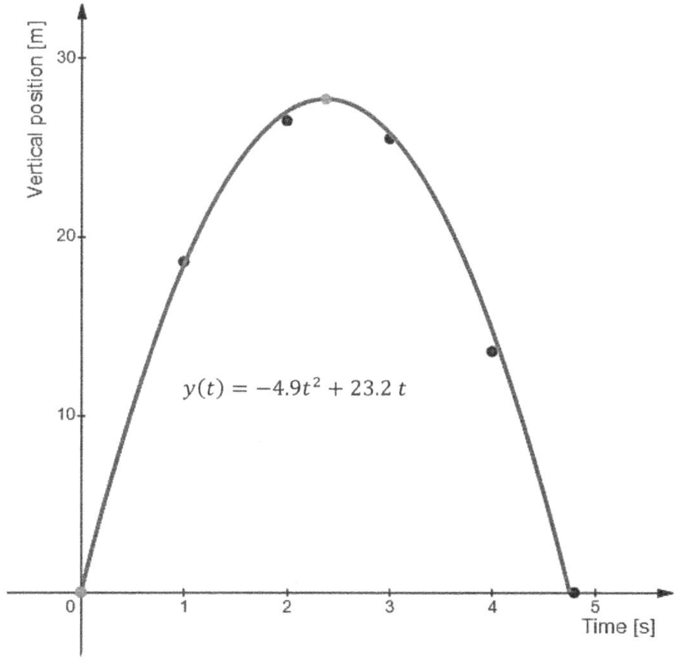

Fig. 8.11 The vertical position of the projected object in the function of time

algebraic description of parametric equations and assure the students that this parameter is shared by both component functions.

$$\begin{aligned} x(t) &= 10.1t \\ y(t) &= -4.9t^2 + 23.2t \qquad 0 \le t \le 4.8 \end{aligned} \tag{8.7}$$

As assessment items often ask students for graphs of velocities and acceleration of a projected object, the students were prompted to derive these functions as well which follows.

8.4.4 *Finding Velocities and Acceleration Functions*

Students who were concurrently taking a calculus class applied the concept of the derivate to formulate velocities and acceleration functions. The rest of the class was prompted to use standard kinematics formulas.

$$\begin{aligned} v_x(t) &= 10.1 \\ v_y(t) &= -9.8t + 23.2 \qquad 0 \le t \le 4.8 \end{aligned} \tag{8.8}$$

Taking the derivative with respect to the time of both equations again lead to

$$\begin{aligned} a_x(t) &= 0 \\ a_y(t) &= -9.8 \qquad 0 \le t \le 4.8 \end{aligned} \tag{8.9}$$

The above equations proved that the object does not accelerate in the horizontal direction but it accelerates in the vertical direction at $-9.8\frac{m}{s^2}$. By mapping up the parameters of the function $y(t) = -4.9t^2 + 23t$ with a general functional form for object's position in a free fall $y = y_o + vt + \frac{at^2}{2}$, the students realized that the initial position $y_o = 0$, the initial vertical velocity $v_o = 23.2$ m/s, and the acceleration is $a = -9.8\frac{m}{s^2}$.

Students were also prompted to formulate the path of motion algebraically to prove its parabolic shape. To accomplish this task, they needed to attend to both position functions. To replace the variables, they were suggested to express the position functions in standard formula notations. Thus considering $x = 10t$, $t = \frac{x}{10}$. Substituting $\frac{x}{10}$ for t in $y = -4.9t^2 + 23t$ resulted in $y(x) = -0.049x^2 + 0.23x$. The function was quadratic, which supported the shape of the path provided earlier in Fig. 8.9.

8.4.5 *Verification Process*

The students were also invited to verify algebraically the trajectory, $y(x)$, for its adherence to its graphical representation. They calculated the maximum position of the projected object using the formula for the vertex of the parabola and the parabola's zeros to find the object's maximum horioznta1 range. The verification process also included derived parametric functions to assure that the object's position calculated using these entities correlated with that on the trajectory picture. They were also to find the resultant velocity at selected instants of time, the maximum height, and the maximum projectile range.

When problem-solving is concerned, other quantities are typically given, yet these examples provided a very comprehensive framework to organize the solution processes. On the next instructional day (not researched in this study), the students used trigonometric ratios to find necessary kinematics parameters and practiced writing parametric equations for provided motion scenarios.

8.5 Treatment Evaluation

The effects of the treatment were to answer the following research problem:

- *Will situating the learning of projectile motion as a two-dimensional motion representation help students realize its mathematical underpinning?*

The treatment's goal was to reflect on the research question of whether highlighting the construction of parametric equations helps students realize distinctive properties of the components of the projectile motion. Thus, supporting students' understanding of projectile motion by considering its path as a combination of two independent position functions was the priority of the enterprise. Stemming from this understanding, another essential property of projectile was expected to emerge; realizing two modes of movements of the projected object, a constant velocity in the horizontal direction and accelerated motion in the vertical one. During the lab activity, the students did not apply trigonometry to resolve motion vectorial parameters into components. This objective was realized on the following instructional units when the students found the components of either initial position, velocity, or acceleration using trigonometric ratios and used their knowledge about the properties of projectile motion developed ealier during the treatment to formulate the parametric functions.

There were two main qualitative-type questions that the students were invited to answer to reflect on the research problem (see Sect. 8.2.3). Students' responses were first categorized into classes based on their common themes and using descriptive statistics the percentages of each category were calculated. The students responded to these questions individually after a week from the date of the treatment. By analyzing the quality of the students' responses, their ability to apply parametric equations was also assessed.

Students' answers were not marked as right or wrong, but these answers allowed inferring how the treatment (labelled as choice (b)) affected their perception of projectile motion when compared to other means they experienced. Table 8.2 shows the percentage of students in each class, and Tables 8.3, 8.4, and 8.5 show verbatim selected responses from each class.

Students (9%, N = 2) who recalled projectile motion observation, choice (a) described the motion qualitatively. However, they could not extract specific quantities to provide legitimate support reflecting on the motion components. The scientific fact that the object does not accelerate in the horizontal direction, which is one of the central kinematics attributes to quantify the motion correctly, did not surface. Reading the students' responses clustered in choice (b) (64%, N = 14), it is reasonable to claim that this constructivist learning environment helped them realize the distinctive motion features of a projected object. The students especially highlighted the linearity of the function in the horizontal direction that proved a constant velocity. Some students mentioned a proven object's acceleration to be $-9.8\ m/s^2$ which is not easily visualized in any experiment.

Choice (c) was selected by six students (27%, N = 6) who referred to the introductory physics courses and mentioned other experiences or labs they performed on

Table 8.2 Students' preference of representation to identify attributes to find the motion components

Response Type	Choice (a) Favored Observation	Choice (b) Favored the Lab Conduct	Choice (c) Favored Other Experiences
Percentage	9% (N = 2)	64% (N = 14)	27% (N = 6)

Table 8.3 Samples students' responses favoring observation: Choice (a)

Student	Response
1	By throwing a projectile. It goes until a max point where it freezes, then starts speeding up down It goes in the ex-direction constantly (minus a little air friction.
2	There is no acceleration in the x-direction based on the ball's usual trajectory when it is parabolically changing its direction toward the y-axis.

Table 8.4 Sample students' responses favoring the lab conduct: Choice (b)

Student	Response
1	When measuring the horizontal distance at each second, the distances were multiples of each other, but the measurements were not proportional; therefore, there was accelerating.
2	We could visibly observe the vertical and horizontal components of the movement in projectile motion. Placing each table for x and y showed a linear graph showing no acceleration in the x-direction.
3	Finding PE enabled me to see the relationship between the motion in both frames of reference. The PE reinforced the idea that the x and y directions each had their functions but with the same parameter time.
4	The most significant piece of evidence about the object's having $-9.8\ m/s^2$ in the vertical direction, and no acceleration in x would be from the parametric lab. The distances and time would change at the same rate showing constant velocity in the x-direction. When the velocity is constant, there cannot be acceleration.
5	The parametric lab showed that while the range increased linearly, the max height could not surpass its quadratic form.

Table 8.5 Sample students' responses favoring other experiments: Choice (c)

Student	Response
1	Pitching a baseball requires you to throw the ball before crossing the horizontal line plane not to hit the dirt because a greater vertical angle yields more velocity.
2	I did a lab when marble rolled up and down in an arched vertical track and then was projected in the air and landed outside of the track (diagram was provided). We had to predict horizontal distance (x) traveled.
3	The monkey is falling from a tree simulator. If a cannon is shot at the monkey as the monkey lets go of the three, the monkey will catch it because it accelerates downward while the projectile takes its path. Therefore, both acceleration and the bullet's acceleration in the y-direction must be the same as the monkey.
4	A lesson I once learned showed the slope of the object shot from a cannon. Vertical velocity was not constant, but horizontal seemed constant.
5	A project about building a trebuchet would fire a pumpkin, and we had to plot the velocity, position, and acceleration for each second in the air.

projectile motion. The labs' contextual richness is unquestionable; however, these students did not recall these labs' specific features that would exemplify the motion's distinctive features when components are concerned.

Question 2 asked to explain the process of analyzing mathematically projectile motion and intended to inform if the students solidified a general mathematical strategy to get ready to solve projectile motion problems.

Debriefing analyses have revealed two categories of answers depending on their scientific validity. Category 1 was marked as showing sufficient understanding that included resolving the motion into components, and Category 2 was marked as showing insufficient understanding that did not provide such prompts. According to the classification, sixteen students (N = 16, 73%) acquired sufficient mathematical background, and six (N = 6, 27%) were still at the developmental stages. Tables 8.6 and 8.7 contain representative responses in each category.

Students in Category 1 provided evidence that they realized the significance of resolving the motion in the x and y components and constructing position and velocity functions, respectively. Resolving the position into components appeared as a necessary step toward analyzing the motion.

Students who did not show sufficient understanding (Category 2) did not explicate the specific motion parameters associated with the kinematics of projectile motion. Their responses missed the precision of describing the motion according to the axis of projection. One can infer from these rather loosely stated descriptions that these students will face challenges to work on projectile motion problems when finding components is necessary.

Table 8.6 Sample responses representing Category 1

Student	Response
1	When projected, an object will reach y-max and x-max. Without a y-component, there is no positive y-movement. Without an x-component, there is no distance in the x-direction
2	This is true because max in the vertical and horizontal planes of the motion acts independently. When the two are combined, they form the trajectory.
3	Visually it is easy to see vertical and horizontal movements, and they are independent. In the lab last week, hitting the tennis ball from the table at different speeds, we proved that motion does not depend on the horizontal velocity.
4	In the labs, finding a separate equation and proof for the horizontal distance and vertical distance traveled. This shows that it is composed of two equations.
5	Projectile motion is the movement of an object through space. The object moves in horizontal and vertical directions.
6	The projectile motion has two completely independent components. The horizontal component has a constant horizontal velocity as it were on a horizontal surface. Moreover, the vertical component accelerates due to gravity as it was falling straight down.
7	Projectile motion is when an object is sent at an initial speed and changes its position vertically and horizontally. These objects have negative acceleration in the y-direction and no acceleration in the x-direction.

Table 8.7 Sample responses representing Category 2

Student	Response
2	The path a projectile takes when fired or thrown
3	The whole motion of one thing plus the velocity, time, and acceleration
4	Projectile motion is when an opposing force acts upon an object with its initial velocity.
5	Projectile motion is a path when an object is projected according to vectors like initial velocity or gravity

Students' abilities to describe projectile motion using mathematical reasoning were considered the critical phase toward handling problem solving that often requires finding components. The majority of these students (N = 16, 73%) seem to acquire a sufficient understanding of the processes that supported the study's hypothesis that highlighting mathematical apparatus helped them build a rich scientific image of this phenomenon. By consistently writing the domain of both functions, the students realized that the time of the motion for both components must be the same. The shared parameter of time, t, for both component functions was to link both motion components into one that constituted the object motion path.

8.6 Summary and Conclusions

While focusing on enriching the analysis of projectile motion by parametric equations, this study reflected on the current recommendations of physics research communities to use the structural domain of mathematics to support the development of science concept's understanding. This study aimed to help students realize that projectile motion is an example of two-dimensional motion that can be resolved into components according to its vectorial representations with a constant velocity in the horizontal direction and varying velocity in the vertical. This understanding was supposed to help students link these ideas to mathematics and formulate parametric functions for any projectile motion. The instrument's evaluation results supported the study objective that the treatment can be considered a promising didactical approach that can help students with two-dimensional motion analyses. It was assumed that the continuity of enhancing traditional kinematics by including parametric equations starting from the first days of studying kinematics generated a better understanding of the distinctive properties of projectile motion. Can curriculum modification suggested in this study result that students will consider the formulation of the parametric equations an integrated part of the two-dimensional analysis? Such a statement is premature. However, this study seems to provide a green light in a direction to accepting this statement.

Physics textbooks (e.g., Giambattista et al. 2007 p.122) provide a stroboscopic photograph of two objects simultaneously falling alongside an edge: one with zeroth horizontal velocity and the other with a non-zero initial horizontal velocity. The stroboscopic pictures provide evidence that the two objects experience the same vertical acceleration and that the time of falling is the same for both objects. While the photographs provide observable evidence for both facts, the pictures do not help realize that each motion's vertical components are identical and they do not provide an insight into the distinctive properties of the motions' components. In the light of developing covariational thinking, this demonstration might not be as beneficial as it is intended to be; the lab includes two objects that in the students' realms generate two separate movements. The question might emerge on what basis the motion parameters of these objects can be compared? What is the mathematical proof for the vertical components of the motion to be the same? Such experiment provides an excellent opportunity for using parametric equations to prove algebraically that the total times of the motion for both objects are the same. This example was brought up to the students. They considered proving the experiment observable outputs as additional evidence of the power of parametric equations to support physics inquiries and make them explicit.

References

College Board. (2015). *Student performance Q&A 2015 AP® Physiscl Free Response questions.*. Retrieved from https://secure-media.collegeboard.org/digitalServices/pdf/ap/ap15_physics1_student_performance_qa.pdf

Clark, R. C., & Mayer, R. E. (2016). *E-learning and the science of instruction: Proven guidelines for consumers and designers of multimedia learning*. Wiley.

Das, C., & Roy, D. (2014). Projectile motion with quadratic damping in a constant gravitational field. *Resonance, 19*(5), 446–465.

Espinoza, F. (2004). Enhancing mechanics learning through cognitively appropriate instruction. *Physics Education, 39*(2), 181.

Gates, P. (2018). The importance of diagrams, graphics, and other visual representations in STEM teaching. In *STEM education in the junior secondary* (pp. 169–196). Springer.

Gamble, R. (1989). Force. *Physics Education, 24*(2), 79–82.

Giambattista, B., Richardson, McCarthy, & Richardson, R. (2007). *College Physics* (2nd ed., p. 122). Mc-Graw Hill Higher Education, Boston. MA.

Jimoyiannis, A., & Komis, V. (2001). Computer simulations in physics teaching and learning: A case study on students' understanding of trajectory motion. *Computers & Education, 36*(2), 183–204.

Jonassen, D. H. (1991). Objectivism versus constructivism: Do we need a new philosophical paradigm? *Educational Technology Research and Development, 39*(3), 5–14.

Jordan, C., Dunn, A., Armstrong, Z., & Adams, W. K. (2018). Projectile Motion Hoop Challenge. *The Physics Teacher, 56*(4), 200–202.

Jörges, B., & López-Moliner, J. (2019). Earth-gravity congruent motion benefits visual gain for parabolic trajectories. *bioRxiv*, 547497.

Klein, P., Gröber, S., Kuhn, J., & Müller, A. (2014). Video analysis of projectile motion using tablet computers as experimental tools. *Physics Education, 49*(1), 37.

Mihas, P., & Gemousakakis, T. (2007). Difficulties that students face with two-dimensional motion. *Physics Education, 42*(2), 163.

Millar, R., & Kragh, W. (1994). Alternative frameworks or context-specific reasoning? Children's ideas about the motion of projectiles. *School Science Review, 75*(272), 27–34.

National Research Council. (2012). *A framework for K-12 science education: Practices, crosscutting concepts, and core ideas*. National Academies Press.

PhET Interactive Simulations. (n.d.). The University of Colorado at Boulder. Retrieved from http://phet.colorado.edu, July 2020.

Prescott, A. E., & Mitchelmore, M. (2005). Teaching projectile motion to eliminate misconceptions. *PME*, 103–109.

Shadish, W. R., Cook, T. D., & Campbell, D. T. (2002). *Experimental and quasi-experimental designs for generalized causal inference*. Houghton Mifflin.

Sokolowski, A., & Capraro, M. M. (2013). Parametrization of motion. *Mediterranean Journal for Research in Mathematics Education, 12*(1–2), 121–133.

Springuel, R. P., Wittmann, M. C., & Thompson, J. R. (2007). Applying clustering to statistical analysis of student reasoning about two-dimensional kinematics. *Physical Review Special Topics-Physics Education Research, 3*(2), 020107.

Wee, L. K., Chew, C., Goh, G. H., Tan, S., & Lee, T. L. (2012). Using tracker as a pedagogical tool for understanding projectile motion. *Physics Education, 47*(4), 448.

Von Glasersfeld, E. (1991). Abstraction, re-presentation, and reflection: An interpretation of experience and Piaget's approach. In *Epistemological foundations of mathematical experience* (pp. 45–67). Springer, New York, NY.

Chapter 9
Reimaging Lens Equation as a Dynamic Representation

9.1 Introduction

Mathematics provides a computational system, reflects a physical idea, or conveniently encodes a quantification rule (Bing & Redish, 2008). A study conducted by Bossé et al. (2010) revealed that the standards and processes in the National Council of Teachers of Mathematics (NCTM's) *Principles and Standards for School Mathematics* (NCTM, 2000) and the five E's—engagement, exploration, explanation, elaboration, and evaluation—developed by the National Research Council (NRC, 2000) display similarities. Thus, a need for more teaching resources emerges. A substantial amount of instruction in mathematics courses is devoted to developing students' skills of graphing and analyzing functions (NCTM, 2000). Students evaluate function limits, determine domain and range, and describe the function behavior using the derivative concepts. Unfortunately, this wealth of techniques of graph analyses is not often applied in sciences. By working on formulas, students' ownerships are reduced to plugging in given quantities and finding a value that makes the resulting equation a true statement. There exist many areas in physics where structural mathematical apparatus can conveniently enrich scientific inquiries. Determining image characteristics using the graph of lens function emerged as a great opportunity of applying the structural domain of mathematics to physics. A more general version of the instructional unit was earlier published (see Sokolowski, 2012). This version is enriched by consistent terminology and a uniform approach to limit interpretation.

A. Sokolowski, *Understanding Physics Using Mathematical Reasoning*,
https://doi.org/10.1007/978-3-030-80205-9_9

9.2 Prompts Used for the Instructional Unit Design

Image characteristics are usually formulated using the *thin lens equation*: $\frac{1}{f}=\frac{1}{d_i}+\frac{1}{d_o}$ and evaluating it for given quantities. Due to a high multiplicity of possible image characteristics—*real* versus *virtual, upright* versus *inverted,* and *diminished* versus *enlarged*—a generalization of these cases is not readily achievable. It is indisputable that the lens equation and magnification formula provide a sound method for computing the characteristics, and a corresponding ray diagram further supports these inferences. However, these processes are applied to conclude isolated cases that do not provide direct access to summarizing image characteristics at its entire spectrum. To help students generalize the phenomenon of image production, a need for using a more general mathematical apparatus emerged. It seemed that converting the lens equation and magnification formula into algebraic functions and investigating the characteristics of the images using corresponding graphs provided a convenient means for enriching lessons about image characteristics. The process revealed that by using the graphs and applying limits, image characteristics were concluded more concisely. Generated graphs opened up opportunities for applying mathematical reasoning normally unreachable by using the traditional lens equation.

The inquiry has two general sections: Sect. 9.2.3 that is about using lens function, and a more detailed Sect. 9.2.4 that is about using magnification function. Considering the complexity of math knowledge needed, in lower level physics courses showing students how to investigate only the lens function would be sufficient. Students of advanced physics courses would also welcome the analysis of the magnification function, which would complete the entire spectrum of image analysis.

In sum, the goal of constructing the instructional unit was twofold (a) to help students with image classification and generalization, and (b) to provide students with opportunities to merge mathematics tools in a pursuit to delve deeper into how natural phenomena behave.

9.2.1 Mathematical Background

It is advised that participants be familiar with sketching rational functions and interpreting asymptotes to blend mathematical and scientific reasoning during this activity. According to *Principles and Standards for School Mathematics* (NCTM, 2000), these cognitive math concepts are primarily studied in precalculus and calculus courses. If learners have an insufficient background, a graphing calculator can assist. While general image characteristics can be concluded from the lens function without applying the concept of limits, it is suggested that students do use limits, e.g., left and right-sided and limits at infinity, and interpret the results in the contexts of the function asymptotes. Limiting case analysis will disclose more details about the image characteristics and show its powerful nature to enrich scientific inquiry.

9.2.2 *Lab Equipment*

This unit can be classified as a guided discovery lesson utilizing an optical bench or a virtual lab. I suggest using a virtual lab because it provides opportunities to verify students' hypotheses conveniently. The virtual lab utilized during the instructional unit is a physics simulation called *Geometric Optics* designed by the Physics Educational Team (PhET) at Colorado University and available for free at http://PhET.colorado.edu/sims/geometric-optics/geometric-optics_en.html. Figure 9.1 illustrates a general setup of the simulation.

Navigation of the simulation is simple, and it allows for exploring various paths of the image production. For this unit, the curvature of the lens is set to be equal to 30 cm. The corresponding focal length of the lens is $f = 15$ cm. The index of refraction is set at $n = 1.4$, and the diameter of the curvature of the lens is 100 cm. Other values of the index of refraction and the focal length are also possible to select. For specific values, the lens might produce images that will be too large to be observed on the computer screen. Thus, a domain for its values, in its mathematical sense, needs to be verified. One parameter calls for more elaboration, the distance of the object. As assigned in the simulation, numerical values of this distance are unparalleled with the functional values in the mathematical sense. The object's distance from the lens is typically considered positive if the object is positioned on the opposite side of the primary focal length. When the functions are graphed in XY coordinates, by convention, positive values are labelled on the right and negative on the

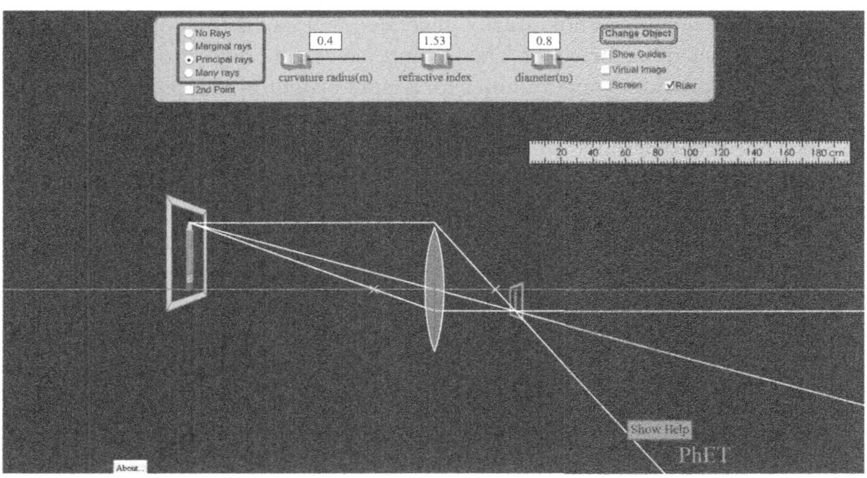

Fig. 9.1 Snapshot of the simulation *Geometric Optics*. (Source: PhET Interactive Simulations (n.d.))

left side of the horizontal axis. The simulation allows placing the object only on the left side of the lens, which might inadvertently suggest negative object's distances. This is not an obstacle, but the instructor might address this nuance to ensure correct scientific interpretations.

9.2.3 Conversion of Lens Equation into a Covariational Relation

This section provides teaching suggestions on converting the lens Eq. (9.1) into a covariational function. An algebraic function, for example, $y = f(x)$, is defined as a relation of two variables—independent x and dependent y—such that for each x value, there is only one corresponding y value (Stewart, 2006). In physics textbooks (e.g., Giancoli, 2005; Tipler and Mosca 2004), the lens formula is called the lens equation and is presented as follows:

$$\frac{1}{f} = \frac{1}{d_i} + \frac{1}{d_o} \tag{9.1}$$

The equation intertwines three different parameters: f, which represents the lens focal length, usually given as a constant; d_i, which represents the distance of the image from the plane of the lens; and d_o, which represents the distance of the object from the plane of the lens. Algebraic interpretation of this formula along with possible explorations are limited. Being derived from the similarity of triangles (Giancoli, 2005), it merely depicts an algebraic entanglement between these quantities that are warranted on the condition of algebraic equity of both its sides. Since it is not expressed as a mathematical function (e.g., using a function notation), its graphing representation is not directly generated, nor can its outcomes be generalized using its functional behavior. The formula needs to be converted into a covariation relation to providing opportunities for a more insightful interpretation that will, in return, offer means for generalizing the physics behind it. To accomplish this task, one needs to determine what scientific covariance is sought. Each constituting parameter can potentially be considered a constant, dependent, or independent. In a typical introductory lab on image analysis (McDermott, 1996), the focal length f is considered a constant parameter, d_i is the parameter of interest of the investigation, and d_o is the independent parameter. Because image characteristics vary as the object's position concerning the plane of the lens varies, it seems convenient to assign the distance of the object as varying independent parameters, the distance of the image as the output parameter, and the focal length as a constant parameter. Such association of the parameters will guide formulating the image characteristics using lens and magnification functions. The equation's parameters will be reimaged to better link and amplify the algebraic entanglement of the quantities with the formality of mathematical notation. This phase will help students to track the covariance between the parameters in their mathematical sense. By replacing the distance of the image, d_i with y, and the distance of the object, d_o with x, Eq. (9.1) takes the following form:

$$\frac{1}{f} = \frac{1}{y} + \frac{1}{x} \tag{9.2}$$

It is to note that the primary parameters of attention are the distances of the object and its images, not their heights. The function notation $f(x)$ was not used due to a possible confusion with the lensfocal length, f. To graph the function and evaluate its limits, the expression needs to be further rearranged to a functional form and illustrate y as a dependent parameter of x that shows Eq. (9.3). Function notation was not used to make a better association with the physical parameters.

$$y = \frac{fx}{x - f} \tag{9.3}$$

Considering the focal length of the lens 15 cm, the algebraic form (9.3) of the rational function that will be used for further analysis takes the following shape

$$y = \frac{15x}{x - 15} \tag{9.4}$$

The instructor highlights the scientific interpretations of the variables; x represents the distance of the object from the lens, and y represents the distance of the image. For coherence of the interpretations, all dimensions will be expressed in centimeters.

The parameter considered a constant is a focal length, f, which can be assigned positive or negative values depending on the type of lens used in the experiment. According to the established convention, a converging lens with a real focal length is denoted by a positive value. A diverging lens having a virtual focal length is characterized by a negative value. In the proposed instructional unit, a converging lens will be used. Teaching suggestions on how to guide the inquiry follows in Sect. 9.2.4.

9.2.4 *Sketching and Scientifically Interpreting the Graph of the Lens Function*

The purpose of constructing the function (9.4) is to learn the *kind* of image, thus *real* or *virtual*, by analyzing a corresponding graph values as positive or negative. While the algebraic function is a concise representation, interpretation of its graph conveniently helps determine the image characteristics based on the object distance from the lens. To activate students' mathematical knowledge, a short review of the

Table 9.1 Summary of image characteristics for a converging lens

Position of the object	Image characteristics
$0 < x < f$	Kind (real/virtual) ________ Orientation (upright/inverted) ________ Magnification (diminished/enlarged/the same size) ________
$x = f$	Kind (real/virtual) ________ Orientation (upright/inverted) ________ Magnification (diminished/enlarged/the same size) ________
$x = 2f$	Kind (real/virtual) ________ Orientation (upright/inverted) ________ Magnification (diminished/enlarged/of the same size) ________
$x \to \infty$	Kind (real/virtual) ________ Orientation (upright/inverted) ________ Magnification (diminished/enlarged/the same size) ________

graphing techniques follows. The students always welcome such a review. As the analysis progressed, students were to complete Table 9.1 that summarized the findings of the investigation.

In mathematics, the process of sketching is supported by identifying the so-called parent function along with its domain and its possible asymptotes that determine the boundaries of the graph. The lens function, $f(x) = \frac{15x}{x-15}$, resembles $f(x) = \frac{g(x)}{k(x)}$, which classifies it as a rational function (Stewart, 2006) and its corresponding graph as a hyperbola. The function domain describes a set of function inputs that produce real outputs; for $y = \frac{15x}{x-15}, x \neq 15, \text{and } x \in R$.

While identifying the domain is sufficient for the mathematical purpose, further restrictions might be imposed due to scientific constraints on the formula from which the lens function was derived. Since x represents the object distance from the plane of the lens and the distance can take only positive values (in the experiment setup), the lens function contains an additional restriction on x that is $x \geq 0$, which results in $x \geq 0$, $x \neq 15$, and $x \in R$. The vertical or horizontal asymptotes will be used to bound the rational function. Subsequent scrutiny to determine these asymptotes for the lens function follows. Function (9.4) possesses two asymptotes, vertical at $x = 15$, due to mathematical restriction on its domain and a horizontal at $y = 15$, due to $y(x \to \infty) = \lim_{x \to \infty} \frac{15x}{x-15} = 15$. The graph of the lens function (Fig. 9.2) is presented to students with its restrictions.

To generalize *the kind* of image characteristics using the graph, it is recommended that students realize that all graphs of lens function with a converging lens (and concave mirrors) will resemble the one illustrated in Fig. 9.2. The difference in appearance will result from choosing focal lengths of different values that affect the position of the function vertical and horizontal asymptotes. Although the positions of vertical asymptotes will exhibit specific variation, all will be located on the positive side of the x-axis. Similarly, all horizontal asymptotes will be located above the

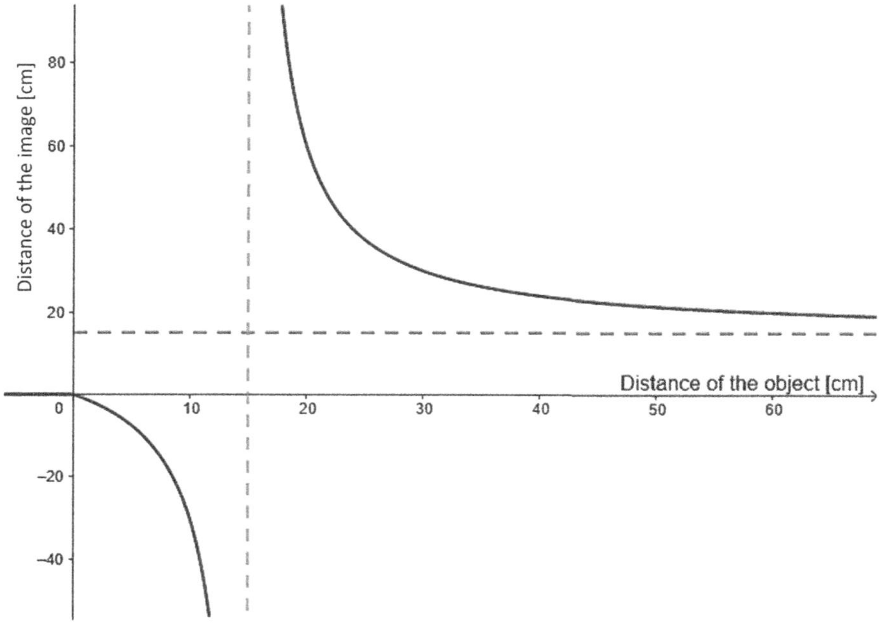

Fig. 9.2 Graph of the lens function $f(x) = \frac{15x}{x-15}$

x-axis. The effects of varying positions of the asymptotes could be supported by graphing technology. Lens function allows determining the kind of the image for specific distances of the object from the lens.

The discussion that follows merges three representations, graphical, algebraic, and scientific. The simulated lab serves as a tool for verification. However, attending to all representations might be overwhelming for the students; therefore, using the virtual lab for verification is suggested while analyzing the magnification function (Sect. 9.2.5). The analysis that follows is organized progressively considering the object's distance from the lens as established in the function domain.

Case 1. The object is placed at x = 0 cm

When the $x = 0$, $f(x) = \frac{(15)(0)}{x-15} = 0$. Thus no image is produced. This conclusion can also be reached by referring to graph Fig. 9.2.

Case 2. The object placed within 0 < x < 15

The y-coordinates of thegraph of the image distance function are negative denoting negative image distances. This graph behavior, according to the general convention of image classification proves that the image is virtual. Realizing the negative values of the distance function comes naturally, referring to the graph. Lens equation produces negative values; however, it does not further support why the distance shows as negative. To align the interpretation to reality, the term *position* would perhaps reflect better on the interpretation instead of distance. Students observe that as the object is being moved closer to the focal point of the lens, the distance of the image increases to infinity, but it remains negative.

Case 3. The object is being moved near the left proximity of x = 15 (x→15−)
To evaluate the function, in this case, the technique of evaluating sided limit is helpful. The notation $y(x\to15^-)$ will denote such condition; $y\left(x\to15^-\right)=\lim\limits_{x\to15^-}\frac{15x}{x-15}=-\infty$. This algebraic statement supports the graph behavior.

Case 4. Object is placed at x = 15 cm
At $x = 15$, the function is undefined, which shows on the graph as a vertical asymptote. Scientifically it means that there is no image produced when the object is positioned at the focal point of the lens.

Case 5. The object is being moved past the focal point, and it is located at the right proximity of x→15+
When the object passes the focal length, but it is still near it, a sided limit $f\left(x\to15^+\right)=\lim\limits_{x\to15^+}\frac{15x}{x-15}=\infty$ helps with the image classification. The result illustrates that the image distance has large and positive values. This function attribute transcends to the discovery that the image is real.

Case 6. The object is being moved away to an infinite distance
This condition also calls for using the limiting case analysis; $f\left(x\to\infty\right)=\lim\limits_{x\to\infty}\frac{15x}{x-15}=15$. The limit value suggest that $f(x\to\infty) = 15\ cm$ which informs that the image is generated at the lens's focal length. A horizontal asymptote $y = 15cm$ of the function supports this conclusion.

Students realize that the distance of the object from the lens and the lens's focal length have a profound effect on the distance of the image and its *kind*. When the object's distance is greater than the focal length, the image is considered real because the function values are positive. The image is virtual when the object distance remains within the lens's focal length because the function values at that domain are negative. A more detailed analysis of the image function will be included in the section that follows. While the *kind* of image is important, its orientation and magnification will complete its optical characteristics. The discussion of the magnification function is more complex; thus, it is recommended for more advanced physics courses.

9.2.5 Formulating Magnification Function

This section advances the analysis, and it will support magnification and orientation of the image. Once the magnification function is derived, the verification process will also take a more sophisticated form; it will incorporate the outcomes of the virtual lab. The starting point of the analysis is the magnification formula

$M = \frac{-d_i}{d_o} = \frac{-y}{x}$. This formula is built of two covariate parameters, the distance of the image, y, and the distance of the object, x. From this form extracting the scientific sense could be a challenge. The magnification function can though be expressed as one variable function, and limiting case analysis can be conviniently applied to learn more about the image.

$$M(x) = \frac{-y}{x} = \frac{-\frac{fx}{x-f}}{x} = \frac{-f}{x-f} = \frac{f}{f-x} \tag{9.5}$$

This analysis will parallel with the one from Sect. 9.2.4. By letting $f = 15$ cm, the magnification function takes the following form:

$$M(x) = \frac{15}{15-x} \tag{9.6}$$

In this equation, $M(x)$ can take positive or negative values, yet it is dimensionless. The positive or negative value of the magnification will help determine the object's orientation, thus *upright* for $M > 0$ or inverted for $M < 0$. The absolute value of the magnification $|M|$ describes the actual size of the magnification. The magnification function $M(x)$ depends only on x, the object's distance from the lens and its graph are presented in Fig. 9.3.

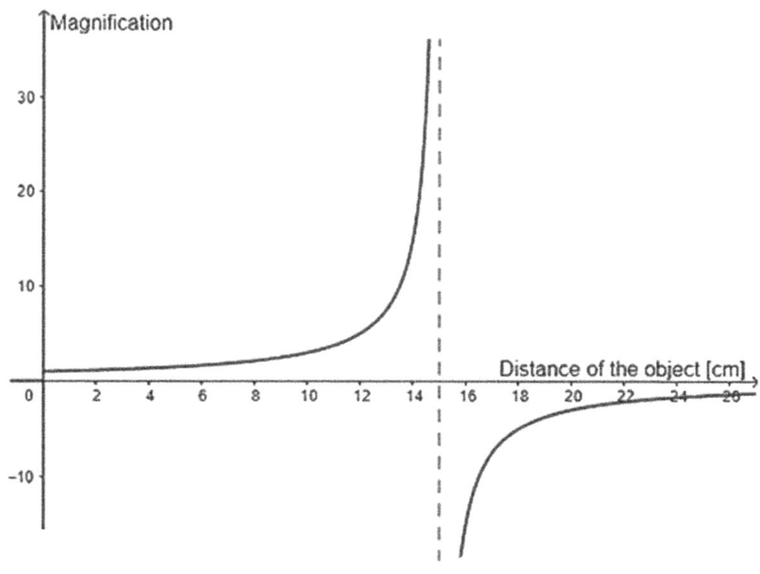

Fig. 9.3 Graph of magnification function. $M(x) = \frac{15}{15-x}$

The graph of $M(x)$ also resembles a rational function. Both graphs' forms are similar, yet the scientific interpretations of these functions' behavior is different because each illustrates different aspects of the image production.

9.2.6 Merging Mathematical and Experimental Representations into One Inquiry

Since the students become acquainted with the interpretation of the lens function, the analysis of the magnification function includes one more representation; the experimental outputs supported by the virtual lab. The instructor might project the simulated lab (see Fig. 9.4) on the classroom screen to verify the conclusions reached from applying mathematical reasoning.

Expressing magnification as one variable function encourages implementing limiting case analysis similarly as it was applied to lens function. The independent parameter of the function represents the object's position concerning the plane of the lens as measured along the optical axis. The analysis will progress analogically to the prior one to highlight similarities of extracting scientific sense from $M(x)$ outputs that will be interpreted as *image orientation* and *magnification*. The interpretation will be consistent with convention found in physics textbooks; positive values of $M(x)$ will imply that the image is *upright*, negative values of $M(x)$ will indicate the image is *inverted*. The magnitudes of $M(x)$ with reference to the value of 1 will classify the image as (a) *diminished* when $0 < |M| < 1$ or (b) *enlarged* when $|M| > 1$. The discussion will begin from the object distance at $x = 0$ cm.

Case 1. The object is placed at x = 0

Physically, it is impossible to place an object at 0 cm from the plane of the lens. Reaching out to the tools of mathematics makes the inquiry possible. The value of

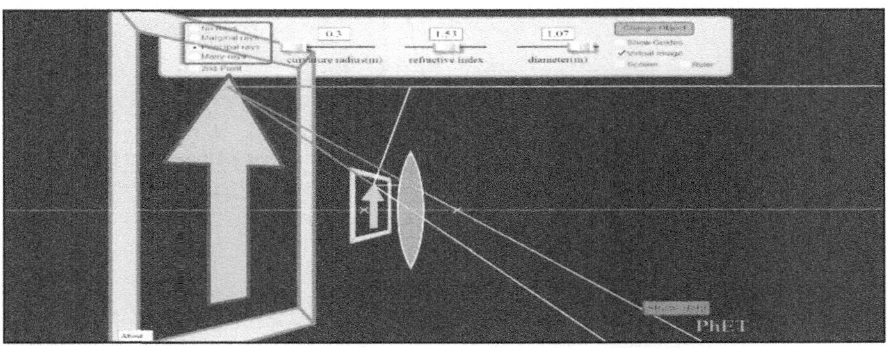

Fig. 9.4 Production of enlarged and upright image. (Source: PhET Interactive Simulations (n.d.))

$M(0) = 1$ appears as the vertical intercept of the magnification graph function on the graph of $M(x)$ (see Fig. 9.3). Magnification of 1 shows that the object and the image have the same heights. The y-intercept of the magnification function illustrates this situation.

This case is interesting; the function $M(x)$ can be evaluated when $x=0$; however, that experimental proof is not possible.

Case 2. The object is being moved from x = 0 to the proximity of x = 15 (0 < x < 15)

This distance interval does not require limiting case analysis because the magnification function can attain real values on that interval, and mathematical reasoning can be induced directly. $M(x)$ attains positive and greater than one value within this interval, which supports the claim that the image is *upright*. According to Fig. 9.3, the function outputs increase from 1 at $x = 0$ to infinitely large values, indicating that the image is *enlarged*. The lens resembles a magnifying glass in this domain. The observation of the image produced by the simulation (see Fig. 9.4) supports the inferences reached from mathematical reasoning. By placing the object close to the lens on its left side and then moving it slowly further to the left, the image appears to be upright and enlarged. By attending to the earlier developed image function, one can add one more characteristic of the image by describing it as *virtual*. The simulation supports this claim because the image is produced on the same side of the lens where the object is located.

Case 3. The object is near the proximity of the focal point on its left side (x→15−)

Evaluation of the sided limits for this case is also consistent with the graph of $M(x)$. When the distance of the image is getting very close to the focal point, thus when $x \to 15^-$ from the left side of the focal point (as seen on the graph function), the image is getting large and the magnification has positive values. This inference can also be concluded from the graph Fig. 9.3, and it can be calculated by evaluating the left-sided limit of the lens function using $M\left(x\right) = \lim\limits_{x \to 15^-} \dfrac{15}{15 - x} = +\infty$. The image is upright because the magnification function takes positive values, and the magnification is very large.

Case 4. The object is placed at the focal point, thus at x = 15 cm

The graph of $M(x)$ displays a lack of corresponding y-values for $x = 15$, which is illustrated in the mathematical sense as a function vertical asymptote for that x value (see Fig. 9.3). Scientifically, this is interpreted as an absence of an image for this specific object's position. The simulation also supports this finding. Due to refracted rays being parallel, no image is produced when the object position is 15 cm (Fig. 9.5).

It is interesting to observe that the corresponding image magnification is also undefined when the object is placed at the focal point, and its algebraic proof

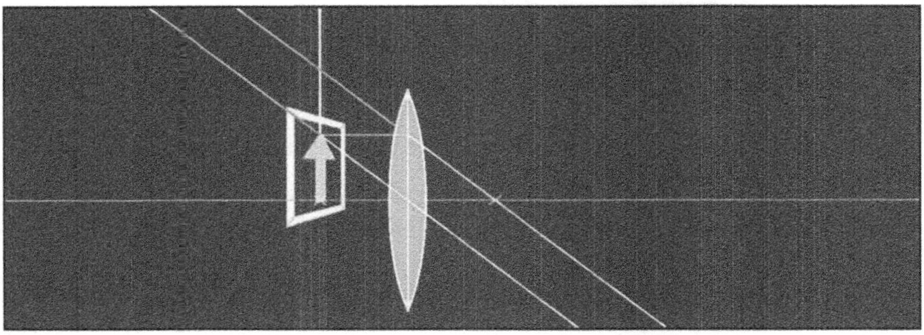

Fig. 9.5 The object is placed at the focal point of the lens, which results in no image being produced. (Source: PhET Interactive Simulations (n.d.))

follows. Since $M(x) = \frac{15}{15 - x}$, and $x = 15$, then $M(15) = \frac{15}{15 - 15}$ is undefined. At $x = 15$, $M(x)$ has also a vertical asymptote; the image is not produced for that x-value and thus the magnification is undefined.

Case 5. The object is moved past the focal point, but it remains at its proximity $x \to 15+$ cm

At $x = 15$, the magnification function, similar to the lens function, is undefined; however, a right-sided limit can be used to describe the object's magnification and orientation when it in the proximity of 15 cm. $M(x \to 15^+) = \lim_{x \to 15^+} \frac{15}{15 - x} = -\infty$ which informs that the magnification as an absolute value of the function values increases to infinity but that the image is inverted. Both the graph and simulation support the mathematical reasoning extracted from the magnification function.

The size of the image using directly the magnification relation, $M = -y/x$, is also very large. The simulation conveniently supports this interpretation (Fig. 9.6).

Case 6. The object is being moved further away from the focal point; $x > 15$ cm

When the object passes the focal length, but it is still near it, $M(x \to 15^+) = \lim_{x \to 15^+} \frac{15}{15 - x} = -\infty$. This result illustrates that the object is inverted and is exceptionally large. As the object is being moved farther away, the image diminishes. This can be concluded from the graph in Fig. 9.3 and the simulation in Fig. 9.7. The image is also inverted due to the negative value of the magnification function. Limiting case analysis can be used to evaluate the magnification function; $M(x \to \infty) = \lim_{x \to \infty} \frac{15}{15 - x} = 0$. The limit result suggests that when the

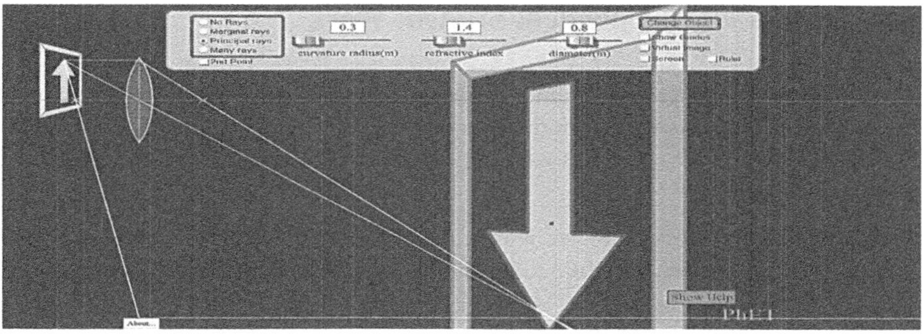

Fig. 9.6 Production of a real image. (Source: PhET Interactive Simulations (n.d.))

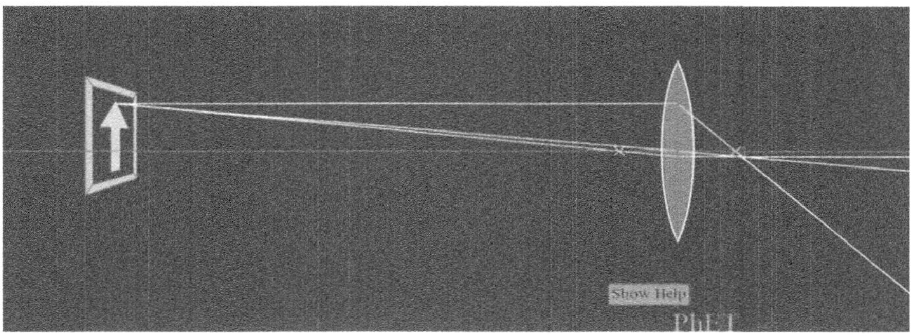

Fig. 9.7 The object is located very far away from the lens. The image is located near the primary focal point shown on the right side of the lens. (Source: PhET Interactive Simulations (n.d.))

distance is considerable, the height of the image is getting exceedingly small. The limit output though does not inform that the height will ever be zero, instead the horizontal asymptote of $y = 0$ places such boundary on its height. Mathematical reasoning can further upraise this result; there is no x-value for which the magnification function would ever be zero. Thus, the image will not disappear; it will exist even though its height will be near the value of zero centimeters (Fig. 9.7).

Analyzing the magnification function concludes the investigations. The structural domain of mathematical and reasoning that stemmed from its application supported the process of generalizing image characteristics that corresponded with images generated by the virtual lab. As the *lens function* helped determine the *kind* of image, *real* versus *virtual*, the magnification function helped delve deeper and describe the image *magnification* and its *orientation*. Coupled interpretation of both functions completed description of the image characteristics.

9.3 Suggested Independent Student Work

The independent students assignment is centered on several objectives, yet similar to the proposed instructional unit discussed herein. Problem 9.1 can also be considered a lab that provides students with opportunities to verify their graph inferences with real lab outputs.

Problem 9.1: Given is a convex lens with a focal length of 20 *cm* and an object initially located remarkably close to the lens, thus $x \to 0$. The object is being moved away from the lens.

(a) Formulate the lens and magnification functions, sketch their graphs and determine the characteristics of the image when the object is located at 10 *cm* and 50 *cm* from the plane of the lens.
(b) Formulate and sketch a lens function and magnification functions if instead of convex lens a concave lens is used and determine characteristics of an image when an object is located at 10 *cm* and 50 *cm*. Did you notice any similarities between the characteritics?

In Problem 9.2 a snapshot of a simulation needs to be provided with a special case of the image produced.

Problem 9.2: The focal length of the lens to generate the images is 40 cm. Write the lens and the magnification function for the image case provided below (Fig. 9.8) and prove algebraically its non-existance.

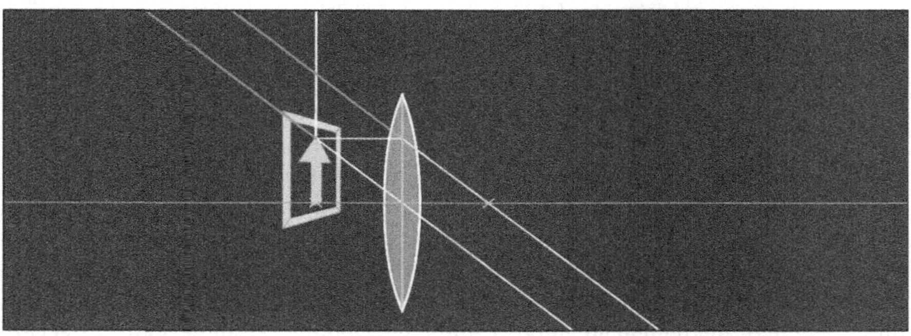

Fig. 9.8 The object is placed at the focal point of the lens. (Source: PhET Interactive Simulations (n.d.))

The context for Problem 9.3 was provided by a graph of a lens function.

Problem 9.3: An image of an arrow is created using a lens. The graph below shows the position of the image in the function of the object's position generated using that lens. All dimensions are given in centimeters.

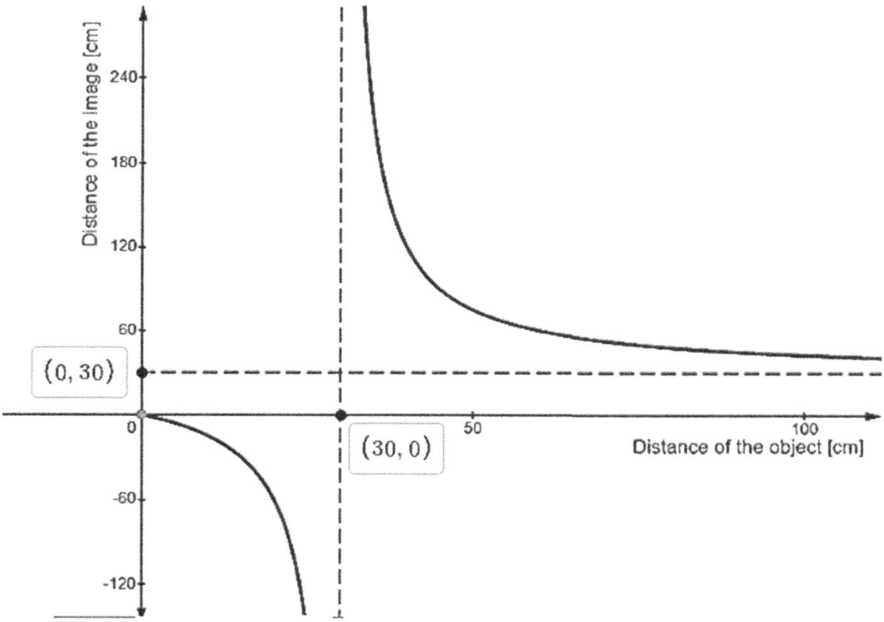

Using the graph, answer the following questions:

(a) What type of lens is used in the experiment?
(b) What is the approximated focal length of the lens?
(c) For what object distances is the image real?
(d) For what object distances is the image upright?
(e) Approximate the image position when the object distance is 10 cm.
(f) What is the magnification of the image when the object is placed at the position of $2f$?
(g) Find the radius of the curvature of the lens used in the experiment.

A graph of magnification function constituted a context for Problem 9.4.

Problem 9.4: The graph below illustrates a magnification function for the experiment described using this graph and the graph from Problem 9.3. (a) Write a lens function for this experiemnt, (b) Determine the characteristics of the image at the indicated object's distances: x=10 cm, x=30 cm, x=50 cm, and for the distance of the object considerably large. Use Table 9.1 to fill in your answers.

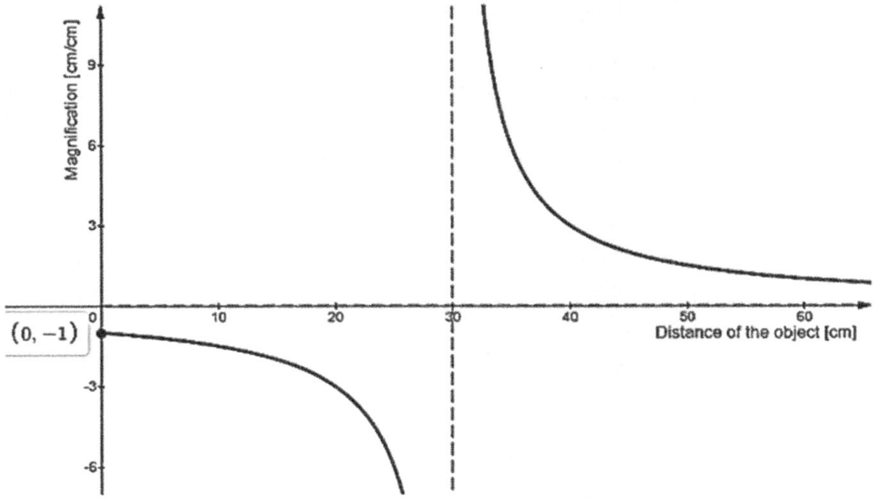

9.4 Summary

This study was not embraced in a formal evaluation. The lens function was though introduced to high school physics students. Students found it exciting to see how algebraic graphs can help generalize physics knowledge. It was concluded that a certain level of maturity and math knowledge is needed to have students fully appreciate such integrated lessons.

While the level of merging mathematical and scientific inquiries in the proposed instructional unit is sufficient to experience supporting role of mathematical reasoning to learn more about natural phenomena, opportunities for extending this inquiry, still exist. The focal length can be interpreted as yet another variable parameter, called perhaps mediating parameter. Such exploration could also constitute an independent student assignment and open up gates for students' to fulfill yet another journey of explorations. The focal length also depends on the lens index of refraction and the lens curvatures, classified as mediating parameters from a physics standpoint. The dependence encompasses these parameters is expressed as $\frac{1}{f}=(n-1)\left(\frac{1}{R_1}+\frac{1}{R_2}\right)$ and is often called the *lens maker's equation* (Giancoli, 2005). More mediating parameters can be identified to gain more insight into the characteristics of the image production and their effects on the image. This stage would depend on the students' physics and mathematical grade level and background. As lens function helps determine the kind of image, *real* versus *virtual*, and magnification function allows describing the size of the magnification and object orientation. Realizing this functional parallelism helps students remember tarditional textbooks rules on determining image characteristics.

References

Bing, T. J., & Redish, E. F. (2008). Symbolic manipulators affect mathematical mindsets. *American Journal of Physics, 76*, 418–424.

Bossé, M. J., Lee, T. D., Swinson, M., & Faulconer, J. (2010). The NCTM process standards and the five *Es* of science: Connecting math and science. *School Science and Mathematics, 110*, 262–276.

Giancoli, D. C. (2005). *Physics: Principles with applications* (6th ed.). Prentice-Hall.

McDermott, L. (1996). *Physics by inquiry: An introduction to physics and the physical sciences* (Vol. 2). Wiley & Sons.

National Council of Teachers of Mathematics (NCTM). (2000). *Principles and standards for school mathematics*. Retrieved from http://www.nctm.org/standards

National Research Council (NRC). (2000). *Inquiry and the national science education: A guide for teaching and learning*. National Research Press.

PhET Interactive Simulations (n.d.) accessed on July 2020.

Sokolowski, A. (2012). Enhancing interpretation of natural phenomena through math apparatus. *International Journal for Mathematics Teaching and Learning, 3*(399), 1–20.

Stewart, J. (2006). *Precalculus: Mathematics for calculus* (5th ed.). Brooks/Cole.

Tipler, P. A., & Mosca, G. P. (2004). *Physics for scientists and engineers* (5th ed.). New York, NY: W.H., Freeman.

Chapter 10
Embracing the Mole Understanding in a Covariate Relation

10.1 Introduction and Prior Research Findings

A diversity of research ranging from a perspective of psychology learning through historical and philosophical views (Furio et al., 2002; Siswaningsih et al., 2017, Feb.) had identified several deficiencies associated with an understanding of the mole; students (a) often identify the mole with mass expressed in kilograms and Avogadro's number of elementary entities; (b) avoid inserting the abbreviation of the amount in the units of moles that results in misinterpretations of the unit, (c) confuse *molar* mass with *molecular* mass. Other studies (Dahsah & Coll, 2007; Musa, 2009) showed that students' weak skills of handling conversion problems involving the mole stem from misinterpreting the meaning of *amount of substance* regarded as a number expressed in terms of moles. Johnstone (1971) found out that students' difficulties with the mole's concept are widespread and are rooted in the complexity of the mole's SI definition. Fang et al. (2014) suggested that "the research of the past 40 years is that the way the mole is conceptualized in educational settings is inconsistent with the meaning of the mole expressed in the SI definition" p. 351.

Attempts to help students understand the concept of the amount of substance as a mole has been made. Uce (2009) used a conceptual change method to develop the mole understanding. Staver and Lumpe (1995) pointed out that the difficulties originated from an insufficient understanding of the conversions' algorithmic routes. Tullberg et al. (1994) claimed that educators teach the mole's concept using extensive reasoning methods instead of direct algebraic equations. Fang et al. (2014) proposed using large rectangular shapes to depict schematically the mole and smaller rectangles inserted inside the larger to represent sub-concepts such as molar mass or atomic mass. Indriyanti and Barke (2017, August) suggested introducing the mole concept by having students count the number of items using weighing, which was employed to realize that the mole's concept was introduced because it

A. Sokolowski, *Understanding Physics Using Mathematical Reasoning*,
https://doi.org/10.1007/978-3-030-80205-9_10

was impossible to count the vast number of particles. Milton (2013) suggested introducing a different, more intuitive, definition of the amount of substance that measured the size of an ensemble of entities (atoms, molecules, ions, electrons, or other specified groups of particles). Erceg et al. (2016) identified several students' misconceptions about the kinetic molecular theory of gases that could be accounted for a weak understanding of units describing the mass of gas.

All the findings targeted various weaknesses of the mole understanding. Each pinpointed a specific aspect of the teaching process rather than suggested a more comprehensive mole introduction method. This study attempts to encompass all these suggestions and especially the one proposed by Scott (2010), who advocated for more integration between algebraic operations and scientific interpretations during the conversion processes. The instructional unit's central theme is to consider proportional reasoning as a backbone for setting up conversion processes and understanding the interpretation of the quantities computed.

10.2 Theoretical Framework

A position has been taken to design instructional support that would explicitly address students' weaknesses. To adequately reflect on the weaknesses, details about the nature of students' misinterpretations of the mole's understanding were summarized. These weaknesses along with suggestions on their elimination are tabularized in Table 10.1.

10.2.1 Weaknesses of the Mole Understanding

Prior research findings and challenges helped to formulate the framework for designing the treatment. The challenges are listed in the left column of Table 10.1, and the actions taken to overcome these challenges constitute the right column of the table.

Recommendations that emerged from the analysis called for unifying algebraic structures to enhance covariational relation between the mass expressed in the mass unit (e.g., kg) and mass expressed in terms of the moles. The review of the algebraic foundations of proportions, rates, and ratios emerged as the priority. Another recommendation targeted the terminology used to verbalize the quantities sought by applying the algebraic structures. For example, instead of *molar mass*, the phrase the mass of one mole of a substance *the mass of one mole of a substance* will be used, at least at the introductory stage of the unit development. This suggestion might seem to have a minor effect; however, when using the phrase *mass of one mole* or *mass of one atom*, the emphasis is placed on the quantity of mass, which students associate with grams or kilograms. It was reasonable to assume that the

Table 10.1 Prompt for formulating the theoretical framework

Challenges with the mole understanding	Suggested interventions
1. Overemphasized conceptual approaches during conversions. 2. High diversity of units used to describe the amount of quantities at a microscopic level. 3. Lack of systematization of the different units of measuring matter.	Use the structure of a proportion throughout various types of conversions systematically. Introduce the mole as a basic unit of the amount of substance in the form; $1mol = 6.02 \times 10^{23}$ *atoms* Introduce *molar mass* before introducing *atomic mass.*
4. Overemphasizing the mole's formal definition as regarded to Avogadro's number and the number of particles in carbon-10. 5. Lack of contrasting traditional units of mass (kg) with the amount of mass expressed in moles.	Introduce the formal definition of the mole after learners are fluent in basic conversion techniques. Provide students with opportunities for contrasting various units of mass and embracing it with conversions.
6. Lack of parallelism to algebraic operations while setting up converting the units of mass, e.g., expressed in kg to moles or a number of particles.	Using the idea of constructing *proportion* and the periodic table of elements to find the reference for molar and atomic masses.
7. Difficulties with interpretations of phrases like *molar mass* or *atomic mass* because the equivalent descriptions (e.g., *kilogram mass* or *pound-mass*) are not used in sciences.	Using the phrase *mass of 1 mole* instead of *molar mass* and *mass of 1 atom* instead of *atomic mass* and consequently apply these statements to build proportions.
8. Diminished effect of algebraic manipulations when applying a detailed dimensional analysis. 9. Lack of connection between atomic mass representing the mass of one mole of a substance.	Replace dimensional analysis by simplifying the units using algebraic algorithms for terms' reduction. Emphasize *mass number* (in the periodic table) as representing the mass, in grams, of 1 mole of a substance.

phrase *molar mass* might unintentionally emphasize the mole as the unit of measurements of the substance expressed in moles.

10.2.2 Proportional Reasoning, Rates, and Ratios

The structure of a proportion and its interpretations will serve as the primary algebraic structure to develop students' algebraic reasoning to help them understand the mole's underpinnings and subsequent conversions. Consequently, an outline of constructing proportions as viewed through the prism of covariation yet enriched to model scientific quantities emerged. While a similar concept called *the laws of proportions* (Mansoor & Rodrigues, 2001) can be found in the literature, this study introduces the learners to the modeling approach of formulating and solving proportions at the start. Departing from the theory of equation, fundamental scientific constants along with their interpretations will follow. Concise structures of mathematics will support quantification processes, and the corresponding scientific embodiments will stem from the outputs of the computations.

Proportional reasoning is a type of covariational reasoning referred to as the capacity to compare the values of two quantities within the same system (Confrey, 2008). The conceptual underpinning of a proportion is usually developed in junior high school (e.g., see Fielding-Wells et al., 2013). Coffield (2000) suggested its more formal introduction at the earlier grade levels to assure its solid conceptual and structural understanding. The applications of proportional reasoning extend to high school and higher-level mathematics. For example, in calculus courses, students formulate differential equations given by a specific entanglement between two quantities or algebraically model dependence on two or several quantities in related rates scenarios. Research has shown that while proportional reasoning is one of the most often applied covariational reasoning, its algebraic underpinnings are better understood with learners' maturity and experience (Clark & Kamii, 1996).

Proportional reasoning denotes reasoning with two variable quantities, y and x, between which there must exist a linear functional relationship of the form $y = mx$, where m represents the proportionality constant often expressed in the form of a constant rate or ratio. The proportionality constant can be picturized as a slope of the linear relationship when sketched. It can also be called a mediating parameter. The functional representation of proportionality is often regarded in mathematics as a *dynamic proportionality* (Miyakawa & Winsløw, 2009). The function can be used to find many values of the dependent or independent variables if a constancy of their rates or ratios within a particular domain is established. The other form of proportionality, more often applied in problem-solving, also used in this study, is called *static proportionality,* which is defined as a specific setup used for computing unique values. Static proportionality intertwines two different phases within a system of continuing the same behavior. Static proportionality or proportion compares two *rates* or two *ratios,* usually with one unknown quantity. What is the difference between ratio and rate?

(a) The ratio is a quotient of two similar or homogeneous quantities with typically the same units, for example, $\frac{10m^2}{6m^2}$, $\frac{1kg}{2kg}$, or $\frac{3L}{8L}$ that when simplified produce dimensionless values.
(b) The rate is a quotient of two different or heterogeneous quantities with different units—for example, $\frac{7m^2}{6s}$, $\frac{20kg}{5m^3}$, or $\frac{3mol}{6L}$. The units of rates cannot be canceled out.

When simplified by applying the division algorithm, ratios and rates represent a *unit* ratio or rate called the *proportionality constant.* For example, $\frac{10m^2}{6m^2} = 2$, $\frac{15kg}{5m^2} = 3\,kg/m^2$. In these processes, ratios might become dimensionless; however, rates retain the units. Furthermore, if the data set satisfies the condition that $y = 0$ if $x = 0$; formulating rates and ratios can establish direct dynamic proportionalities called linear functions. How to convert a static proportionality into a dynamic one or vice versa? Consider a linear functional relationship $y = mx$ and a direct proportionality constant expressed as a slope $m = \frac{y_1 - 0}{x_1 - 0} = \frac{y_1}{x_1}$. Substituting $\frac{y_1}{x_1}$ for m in $y = mx$ results in

$$y = \frac{y_1}{x_1} x \tag{10.1}$$

Equation (10.1) represents a dynamic proportionality allowing to compute y or x when either of these quantities value is provided and be sketched as a linear graph (Fig. 10.1). Suppose that the mass of one mole of magnesium is concerned. This value can be represented as a slope $m = 24\frac{g}{mol}$ and embraced in a dynamic proportional relation, $y = 24\frac{g}{mol}x$ that allows computing either the mass, y for a given number of moles of magnesium, x, or vice versa. This direct proportionality relation can be displayed in XY coordinates, with labeling corresponding to the quantities that it models.

The function inclination to the horizontal axis is the slope that can also be computed using $m = slope = \frac{rise}{run}$ with a value of m = 24. Sketching more graphs for different substances in the same XY axes would conveniently provide means for discussing their relative molar masses and chemical properties.

How to convert a dynamic proportionality into a static one? Static proportionality does not retain covariate relationships; instead, it refers to its selected values.

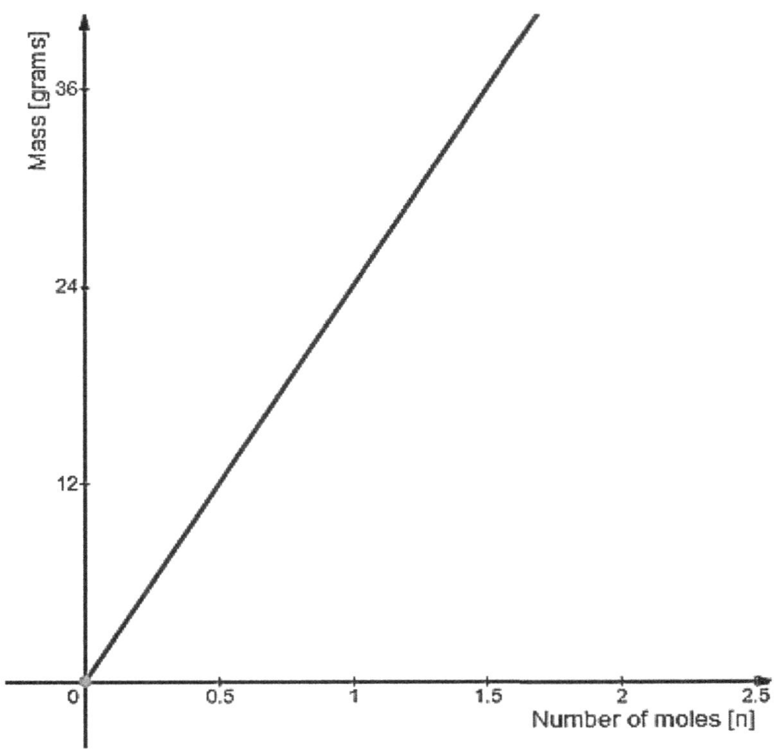

Fig. 10.1 Graph of mass versus the number of moles for magnesium

Replacing y by y_2 and x by x_2 in Eq. 10.1 results in $y_2 = \frac{y_1}{x_1} x_2$. Expressing the terms as ratios produces a static proportionality:

$$\frac{y_2}{x_2} = \frac{y_1}{x_1} \tag{10.2}$$

By applying a cross multiplication and division, this proportion can be converted to a proportion of two ratios, respectively

$$\frac{y_2}{y_1} = \frac{x_2}{x_1} \tag{10.3}$$

A static proportion, either containing two rates or ratios, can be formulated only for quantities satisfying the condition of a *direct proportionality*. For example, the proportions; $\frac{20kg}{5m^3} = \frac{x}{4m^3}$, or $\frac{4m^3}{5m^3} = \frac{x}{20kg}$ represent direct proportionality because, from an algebra point of view, a corresponding dynamic representation will lead to a linear function with the initial value of (0,0); $y = mx$. From a scientific point of view, the condition informs that e.g., if the object's volume is zero, its mass must also be zero. It is usually up to the students to decide whether they use an equivalence of two rates or two ratios to construct a proportion and solve it. While literature does not provide evidence on which of these proportionalities is more accessible to students, a proportionality of two ratios (Eq. 10.3) will dominate in this study.

Understanding proportionality extends beyond setting up proportions and computing the value of the missing quantity. It is linked with one of the most fundamental functions in algebra, a linear function that supports understanding other more complex functions. Realizing its high scope of applications as well as limitations will intertwine with scientific underpinnings during the treatment.

10.3 Methods

This study was conducted with 25 (10 females and 15 males) freshman college physics students who did have a prior background about the mole and its conversion from high school chemistry and physics courses. Most of the students (N = 18) majored in engineering programs, and six of them (N = 6) majored in other programs. Most of these students (N = 20, 80%) were Caucasian, and the rest (N = 5, 20%) consisted of other races. The students graduated from various high schools, and all took part in the study voluntarily.

This undertaking can be classified as a pretest-posttest one-group experimental study (Shadish et al., 2002) with predominantly qualitative analysis of its results. The instructional unit was developed to address students' deficiencies in handling the mole understanding listed in Table 10.1 and its subsequent conversions. The treatment design also took into account the results of the pretest taken by this study's participants. The purpose of the pretest was bifocal; it served (a) as an additional source of prompts to design the pedagogy of the instructional unit (b) to assess students' ability to adapt the proportional reasoning techniques to make conversions. The following question guided this study:

> *Will proportional reasoning help students conceptualized the meaning of one mole and subsequent mass conversions to the amount of substance and vice versa?*

This study was also to determine if the students could adopt different computational techniques for conversion processes.

10.4 The Lecture Component

The instructional unit was designed for one class period (55 min), and it consisted of several parts organized deductively. Thus, it began with more general conversions and zoomed gradually into more detailed ones. Each part was initiated by conceptual questions whose purpose was to activate students' proportional reasoning before using formal algebraic conducts. The instructional unit took the form of a discovery-type lesson. The instructor posited inquiries, let students discuss the answers, and then the corrected conversion process was applied to prove or disprove students' claims. While the unit could be narrowed to conceptualizing the mole's idea, a more pragmatic approach was taken to also include the mole's conceptualization as an amount of substance and merge it with developing conversion techniques. It was hypothesized that providing students with opportunities to illustrate a higher diversity of mass conversions and offer techniques for handling them will provide tangible learning experiences that students could apply in their other science courses.

10.4.1 The Mole as a Fundamental Unit of the Substance Amount

From a scientific point of view, the mole's idea might be challenging because it does not contain the traditional mass units such as kilograms or grams that students are trained to applying, thus addressing this nuance beforehand can help. The lecture was initiated by stating that on a macrolevel, a sample of a known substance can be characterized by its mass (in kg). In contrast, the sample can be described on a microscopic level by the number of its entities. These entities are atoms whose samples of milligrams order, or less, are needed for microscopic experiments. How to handle counting the atoms? Suppose one needs 10^{30} atoms. Counting atoms as single entities is impossible, and a unit representing a more substantial amount was needed. The unit, convenient for calculations and laboratory practice, consisting of 6.02×10^{23} atoms was called *the mole*, and the base conceptualization of the quantity is called the *amount of substance,* not the amount of mass. If there are 6.02×10^{23} atoms in a container, then one can claim one mole of the atoms in the container. The students were assured that a more formal definition of the unit of one mole would be provided later. The teacher also highlighted that while the *amount of mass (not substance)* must be expressed in the units of mass (e.g., grams or kilograms), the phrase *amount of substance* does not contain the term *mass*. Thus, it does not induce the units of grams or kilograms. In sum, the mole's unit does not explicitly illustrate the amount of mass traditionally expressed in kilograms. Consequently, it does not account for the mass of one atom. However, subsequent algebraic operations allow various conversion processes. While the mass of a substance is usually denoted by a lower-case *m,* e.g., $m = 4\ kg$, the amount of substance is represented by *n*, e.g., $n = 10$ *moles* that is abbreviated to $n = 10\ mol$.

To further conceptualize the mole's idea, the teacher informed that as most fundamental physical quantities are measured using devices, e.g., temperature by a thermometer or electric current by an ammeter, no instrument measures the amount of substance (in moles) directly. However, by applying algebraic operations, a substance mass measured by a balance scale expressed in kilograms can be converted to moles. The ultimate question emerged how to convert the mass of a substance to a corresponding number of moles? Before immersing in mass-mole conversions, the instructor introduced a technique of converting a number of particles to moles to support conceptual understanding of the concept of one mole and to highlight the mechanism of applying proportional reasoning. These conversions are intended to have students realize that the mole's unit can link the micro - and macro - worlds and *vice* versa.

10.4.2 Converting the Number of Atoms to the Units of Moles

The teacher posited the following question: Does the number of moles of a substance depend on the number of atoms? After a short discussion confirming that it does not, he posited the following problem.

Example 10.1: Convert 10.04×10^{23} atoms of silver to moles.

To convert the number of atoms to moles, the teacher assigned the variable of interest as *n* represented by 10.04×10^{23} *atoms* and demonstrated the process of formulating two parallel statements that will lead to making up a proportion; first containing fundamental constant called in this study *a known (conversion) statement*; $1mol = 6.02 \times 10^{23}$ *atoms* and the other containing the variable of interest;

$$Known\,statement: 1mol = 6.02 \times 10^{23}\ atoms$$

$$The\,statement\,with\,a\,variable: n\,moles = 10.04 \times 10^{23}\ atoms$$

While to solve for n, the above setup of equations involved two algebraic operations instead of one if the *statement with a variable* were placed in the first line, this arrangement was purposeful and it was to highlight the mechanism of the proportional reasoning. Indeed the construction of all the proportions during this lecture was initiated from writing down the *known statements*. From an algebra point of view, the statements' order does not affect the final answer, and either is fine. The parallelism in formulating these statements is worthy of highlighting though; both contain moles on their left sides and atoms on their right sides, and the equal sign is vertically aligned. Such alignment is easier for students to remember. By embracing the statements in the algorithm of division, a proportion of two ratios emerged.

$$\frac{1}{mol} = \frac{6.02 \times 10^{23}\ atoms}{10.04 \times 10^{23}\ atoms} \tag{10.4}$$

By cross-multiplying, canceling the units of atoms, and solving the proportion for n, the students learned that

$$n = \frac{(1mol)(10.04 \times 10^{23}\ \cancel{atoms})}{6.02\ \times 10^{23}\ \cancel{atoms}} = 2\ mol \qquad 10.5$$

Some students had figured out the answer before solving the proportion. Nevertheless, the purpose of writing the statements was to have them understand the mechanism of setting up the equations whose general structures will be used for other types of conversions. Initiating the proportion using fundamental statements rather than ratios was to have students realize the underlying foundation of formulating the equity of the sides of the proportions. To enhance the understanding of the mole as a quantity linking the macro-world with the micro-world, the teacher posited a further question; can the amount of any substance expressed in moles be converted to atoms? After a short discussion, he wrote Example 10.2 on the board.

Example 10.2: How many atoms of monoatomic hydrogen are there in 4.2 moles of the substance?

The instructor pointed out that to solve the problem it is advisable to use a similar structure for formulating a proportion. Let x represent the number of atoms in 4.2 moles. The teacher did not use n to label the number of atoms to avoid confusion with n denoting the number of moles. The consistency happened to be unnecessary as almost all students who applied proportions on the posttest used x to denote any variable of interest which was fine.

$$Known\ statement : 1mol = 6.02 \times 10^{23}\ atoms$$

$$The\ statement\ with\ a\ variable : 4.2\ moles = x\ atoms$$

These statements lead to the following proportion:

$$\frac{1\,mol}{4.2\,mol} = \frac{6.02 \times 10^{23}\, atoms}{x\, atoms} \quad (10.6)$$

When solved, $x = 25.28 \times 10^{23}$ atoms.

After solving a couple of more examples of a similar conceptual type, the teacher invited the students to determine if the kind of substance expressed in a number of atoms affected the corresponding amount expressed in moles. By observing the patterns of formulating the proportions, students realized independence of the properties of the unit of the mole from the atomic mass of the substance.

Students noted that mole-atoms conversions could be embraced in a dynamic proportion; $\frac{1 mole}{n} = \frac{6.02 \times 10^{23}\, atoms}{x}$ that lead to its more general version applied in the dimensional analysis; $x = \left(6.02 \times 10^{23} \frac{atoms}{mole}\right) n$, where x represented the number of atoms and n, the number of moles. The shortcut stemmed from a static proportion, it was not given as a statement. Frequent use of 6.02×10^{23} *atoms* prompted an introduction of Avogadro's number. After assuring that the students realized the mole's unit as a designated number of atoms and were comfortable with mole-atoms conversion, the instructor proceeded to the second phase that dealt with the mass-mole conversion.

10.4.3 Converting Mass of Substance to Moles

This type of operation dominates in science, and it involves a reference to the periodic table of elements. Due to a low percentage of correct answers on the pretest (32%), (see Sect. 10.5.1 and suggestions of the prior research), this conversion path required a more detailed approach. To alert the students that if a substance mass in kilograms is given, then the number of moles will depend on the substance's atomic mass, the teacher posited the following conceptual problem. The problem will be solved after discussing with the students all involved lines of linking the quantities interpretations.

Example 10.3: Does 2 kg of mercury $^{200.59}_{80}Hg$ contain the same number of moles as 2 kg of potassium, $^{39.102}_{19}K$? Is it necessary to use Avogadro's number while converting these masses to the number of moles?

Most of the students claimed that the number of moles would not be the same and suggested that Avogadro's number be included in the proportion along with the mass of one atom of each of these substances. The students realized that when the mass in kilograms is given, a covariate relation of one atom's mass must be used first to find the number of atoms. Following this line of thought, the number of moles could be seen by setting up a proportion like in Example 10.1. Overusing Avogadro's number was coded by the prior research as students' *misinterpretations* leading them to incorrect conclusions regarding conversion. Thus, while this approach had its merit, the students realized that the Avogadro's number would have to be used twice in two inverse algebraic operations, multiplication and division, which would yield an identity operation. The teacher redirected the students' thinking toward interpreting the *mass number* as a mass, expressed in kg, of one mole of a substance. He reviewed the interpretations of numbers associated with atoms as described in the periodic table of elements; ${}^{A}_{M}X$ where A is called *the mass number*, and M (sometimes labeled as Z) is called the *atomic number*. As given in the periodic table of elements, the mass number is expressed in terms of u, the atomic mass unit that equals 1.66×10^{-27} kg. In conversion problems, the *mass number* can take a dual interpretation (see Fig. 10.2). It can be interpreted as a *mass of one mole of atoms* of a substance (also called *molar mass* and expressed in grams) or a *mass of one atom* of the substance (also called *atomic mass)* and expressed in kilograms if multiplied by the atomic mass unit u).

Both performances are mutually dependent, and one can be converted into the other or used as indicated by a specific task. Understanding the differences between these two different yet mutually coherent interpretations will be exemplified in the forthcoming conversion problems. The teacher informed the students that they would use the mass number as representing the mass in grams of one mole of a substance during this phase of inquiry. Realizing the differences in formulating proportion when finding the number of moles of a substance instead of finding the mass of a certain amount of substance moles was highlighted by the instructor as a critical determinant of setting up the proportion statements.

While in mathematics, setting up proportion is determined often by the symbols of variables, in science, correct interpretations of the tasks also contribute to correctly formulating algebraic statements. After that inclusion, the teacher referred to the problem stated and pointed out the elements' positions in the periodic table while asking the students to predict the answer before applying the formal process. The predicting phase was to have students use their mental image of the idea of 1mole, which, when coupled with the interpretation of notations used in the periodic table, provided a bridge to traditional interpretation of mass units. Students were ready now to construct proportions and convert both substances to moles. Example 10.3 is now converted to a task to prove.

Example 10.4: Prove that 2 kg of mercury, ${}^{200.59}_{80}\mathrm{Hg}$, contains a different amount of moles than 2 kg of potassium, ${}^{39.102}_{19}\mathrm{K}$.

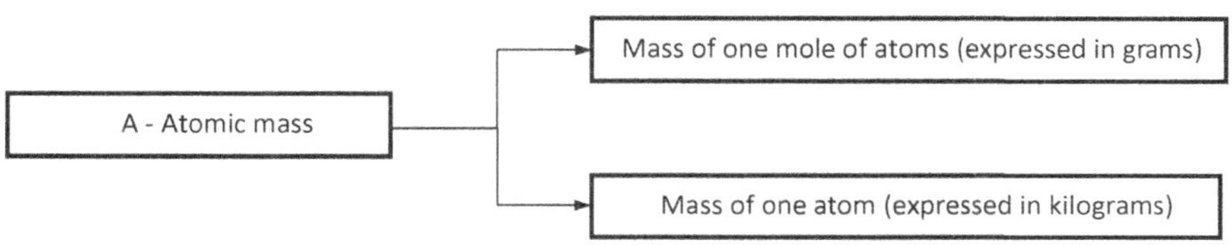

Fig. 10.2 Dual interpretation of the atomic mass number suggested in conversions

This problem requires setting up two independent proportions, finding the number of moles in 2 kg of mercury and finding the number of moles in 2 kg of potassium. Referring tc the established in Example 10.1 structure for enacting a proportion, the teacher labeled n as representing the number of moles in 2 kg of mercury.

$$Known\,statement : 1mol = 200.59g$$

$$The\,statement\,with\,a\,variable : n\,mol = 2000g$$

$$\frac{1mol}{n\,mol} = \frac{200.59g}{2000g} \tag{10.7}$$

Thus $n = 9.97\ moles$. By labeling the variable and setting up a similar proportion for potassium, the students learned that for potassium, $^{30.102}_{19}K$, $n = 66.44\ mol.$ Students confirmed that their predictions were correct that potassium with a lower mass of one mole of the substance generated more moles. When given by the number of atoms, the number of moles is independent of the substance type. However, the number of moles depends on the kind of substance when the mass of one atom characterizes the substance.

10.4.4 Converting Mass of a Substance to the Number of Atoms

This type of operation was designed to find conceptual connections between Examples 10.1 and 10.2 and practice using the mass number as representing the mass of one atom.

Example 10.5: How to set up a proportion to find the number of atoms in $4kg$ of copper?

Following the structures of previous conversions, the students suggested applying two steps to answer the question; first, to convert the mass to moles (as in Example 10.2) and then by using the property that $1mol = 6.02 \times 10^{23}\ atoms$ (as in Example 10.1) find the number of atoms. This path of thinking illustrated their conceptual understanding of the fundamental conversion techniques. After confirming the answer, the teacher intended to expand the type of conversions and suggested using the mass number as a mass of one atom of the substance expressed in kilograms. By the periodic table of elements, copper is described as $^{63.54}_{29}Cu$. How to construct a proportion that would use the known statement of the mass of one atom of copper?

Let x represent the mass of one atom of copper.

$$Known\,statement : 1\,AMU = 1.67 \times 10^{-27}\,\text{kg}$$

$$The\,statement\,with\,a\,variable : 63.54\,AMU = x\,kg$$

Creating a proportion and solving it for the variable results in $x = 1.06 \times 10^{-25}\ kg$ that represented the mass of one atom of copper. This process did not provide the answer for the initial question yet, and another proportion was needed. Let x represent the number of atoms in 4 kg of copper.

$$Known\ statement: 1\ atom = 1.06 \times 10^{-25}\ \text{kg}$$

$$The\ statement\ with\ a\ variable: x\ atoms = 4\ \text{kg}$$

By solving the proportion, one learned that $x = 3.77 \times 10^{25}$ *atoms* represent the number of atoms in 4 kg of copper. While the process depicted all the details, most students simplified it by formulating the proportion $\frac{1}{x} = \frac{1.06 \times 10^{-25}\ kg}{4kg}$ or just writing the final step $x = \frac{4kg}{1.06 \times 10^{-25}\ kg}$, which was also accepted.

After solving these examples, the students were invited to solve textbook problems and design a theoretical experiment that would formulate a regression line whose slope would represent the mole's unit. This task aimed to link static and dynamic representations of proportions and have the students become more fluent in translating between these representations. During the following lecture that is not discussed in this study, the instructor introduced a formal definition of one mole. He also made further connections to Avogadro's number and carbon -10. To bring forth applications, he used the conversion techniques to solve problems regarding ideal gas law; $pV = nRT$ and $pV = Nk_BT$, where p represented pressure inside the gas expressed in Pascal's, V volume of the gas expressed in meters cubed, n number of moles, $R = 8.31 \frac{J}{mol} \bullet K$ gas constant, T gas temperature expressed in Kelvin scale, N number of atoms, and $k_B = 1.381 \times 10^{-23} J/K$ Boltzmann's constant. While working on converting, students were given options of using any methods of their preference.

10.5 Pretest Posttest Analysis

The study used qualitative analysis of students' responses, which reflected on its aim. Statistical descriptive analysis was used to assess True/False multiple types of questions.

10.5.1 Analysis of the Pretest Results

Three questions, designed by the author and consulted with science professionals, were used to assess the students' understanding of the mole's unit and its conversion. Item #1 was a True/False type, item #2 required the students to convert the given mass to moles, and item #3 assessed students' level of understanding of the

Table 10.2 Students' pretest responses (Item #1, True/False)

Multiple choices	Answer/percent correct
(a) The mass of two moles of oxygen is the same as the mass of two moles of nitrogen.	False, 88% (N = 22)
(b) Two moles of iron contain the same number of atoms as two moles of zinc.	True, 56% (N = 14)
(c) Molar mass can be expressed in kilograms or grams.	True, 32% (N = 8)
(d) Both atomic mass and molar mass can be used to find the mass of a certain number of particles.	True, 72% (N = 18)

definitions of: 1 mole, molar, and atomic mass. Table 10.2 illustrates the descriptive analysis of item #1 and the evaluation of the remaining items follows.

The percent of correct answers on Item #1, part #a, was relatively high (88%), which showed that the students perceived the mole unit as the quantity measuring the amount of substance not mass. The percent of correct answers on the remaining parts of this assessment item varied. A relatively low percentage of the correct answers (32%) was reported on the molar mass interpretation (c). This deficiency was addressed while formulating the theoretical framework.

Item #2 asked the students to convert 2 kg of sodium $^{22.99}_{11}Na$ to moles. It was correctly solved by 20% (N = 5) of the students. About 24% (N = 6) of the students did not attempt to solve it, 15% (N = 4) got lost in setting up the conversion process (while using the dimensional analysis). The remaining 39% (N = 10) inserted the Avogadro's number in the dimensional analysis and could not carry out the units' cancellation. None of these students attempted to apply proportional reasoning to solve the problem.

Item #3 asked students to define the mole's unit, molar mass, and atomic mass using their own words. The analysis showed that the students understood the mole's conceptual definition; most of them (80%, N = 20) stated that the mole represents the number of particles/atoms/entities a substance has or an object is made of. The remaining 20% (N = 5) used general descriptions. Verbatim, for example: "*It is used to describe matter and can be converted to other things*," "*a unit of measure to express mass, volume, and particles*," "*a basic unit used for mass.*" The students could not conceptualize the phrases; *molar mass* and *atomic mass* on which the percentages of correct answers were respectively (25%, N = 6) and (30%, N = 8). The pretest questions were not returned to students, nor were they discussed until the study was completed.

10.5.2 Comparisons of the Pretest and Posttest Results

The students took the posttest after about 6 weeks from the date they took the pretest. The format of the posttest questions was like the one given on the pretest. A summary of the comparison of pretest/posttest responses to Question 1 is included in Table 10.3. The percentages show the correct answers.

Table 10.3 Students' pretests and posttest responses to Question #1

Question: classify as true/false	Pretest	Posttest
(a) The mass of two moles of oxygen is the same (b) As the mass of two moles of nitrogen	False, 88% (N = 22)	90% (N = 23)
(c) Two moles of iron contain the same number of atoms as two moles of zinc.	True, 56% (N = 14%)	85% (N = 21)
(d) Molar mass can be expressed in kilograms or grams.	True, 32% (N = 8)	78% (N = 20)
(e) Both atomic mass and molar mass can be used to find the mass of a certain number of particles.	True, 72% (N = 18%)	96% (N = 24)

The posttest results indicate an improvement in all areas tested, especially in interpreting *molar mass* (c). This improvement might be accounted for highlighting established verbal definition as *the mass of one mole* and suggest that the commonly used phrase *molar mass* does not explicitly convey the quantity's physical unit and its correct interpretation to students. At a similar rate (80%, N = 20), the students succeeded in converting 2 kg of sodium, ${}^{22.99}_{11}Na$, to moles. The results on both questions supported the hypothesis that using the structure of proportion and embracing the process with consistent and systematic reasoning helped students understand the interpretation of the mole's unit, learn, and apply the conversion techniques. Percentages of students who claim that both *molar mass* (86%, N = 22) and *atomic mass* (90%, N = 23) can be used to find a mass of particles were also higher on the posttest.

Several interesting inferences emerged from a more detailed descriptive analysis. While initially, I assumed that knowing the percent of correct/incorrect answers would assess the intervention, I decided to zoom more in-depth into how students solved the problems and evaluate their methods. Out of the 20 students who correctly converted 2 *kg* of sodium to the number to moles, (20%, N = 4) applied dimensional analysis, and 80% (N = 16) constructed proportions. On the pretest, no student used proportions in the conversion process. It is to mention that these students were free to use any method of their choice to answer these questions. While on the pretest, two of these students attempted to apply the dimensional analysis; these students adopted the method of proportions on the posttest.

Zooming deeper, it was interesting to learn what caused the remaining five students not to succeed in conversion problems. Two of the students did not attempt to solve it while having the proportion set up correctly; two included Avogadro's number in the dimensional analysis and got lost during the conversion process. One student incorrectly converted kilograms to grams, which resulted in incorrect mole computations. A strong association of the mole's unit to Avogadro's number still prevailed in these students' minds even though Avogadro's number was not emphasized during the instructional unit. Misuse of the Avogadro's number was also reported by other researchers (e.g., Furio et al., 2002).

The percentages of students who correctly interpreted (a) the unit of the mole were (N = 24, 96%), (b) the molar mass (86%, N = 22), and (c) atomic mass (90%,

N = 23). The percentages were higher on the posttest when compared to the pretest. It is believed that bringing forth students' mathematical reasoning in a manner consistent with what they had studied in their mathematics courses allowed the transitioning of these skills and supported the conversion processes.

10.6 Summary and Conclusions

The purpose of this study was to propose alternative ways of teaching the conversions of micromacro worlds by using proportional reasoning, fundamental constants, and the core idea of substance amount expressed in the units of moles. While the study's focus was to develop conversion techniques, understanding the idea of molar mass and atomic mass was also considered. I infer that augmenting molar and atomic mass verbal descriptions helped students understand the units' meaning and set up correctly the algorithms for conversions. Furthermore, consistent integration of mathematical structures with the scientific interpretation recommended by (Niaz, 1989; Tullberg et al., 1994; Scott, 2010) also had its high contributions. While the prior research suggested such integration, this study attempted to provide more in-depth didactical approaches readily available to be duplicated. It is somewhat misleading to assume that students will figure out how to integrate algebra rules during scientific quantification processes by themselves. The integration of both subjects' contents must be developed using involved contexts and delivered with attention to all details, algebraic and scientific so that students can realize their underlying mutual unity and adopt these methods in school practice.

It is reasonable to suggest that providing students with methods whose algebraic structures are familiar to students can improve the mole's understanding and other computations not necessarily related to conversions.

While using more pretest/posttest items and assessing the intervention from a wider angle would provide means for a more comprehensive evaluation of the intervention instrument, it is believed that the quantitative/qualitative nature of the items provided sound evidence of its didactical potential. One of the limitations of the study was the relatively low sample size of the group under treatment. However, it is believed that curriculum policymakers and other stakeholders can consider the encouraging posttest results to adopt the didactical merit of the unit in physics, chemistry, and lower levels of science courses.

I hope that this study will initiate further research to assess students' understanding of other scientific concepts where proportional reasoning can be applied. Furthermore, unifying the methods of teaching the mole across physics and chemistry and other science courses would not only help solidify the approaches but also encourage instructors and students to use covariate reasoning to unify other inquiries in science.

References

Clark, F., & Kamii, C. (1996). Identification of multiplicative thinking in children in grades 1–5. *Journal for Research in Mathematics Education, 27*(1), 41–51.

Coffield, F. (Ed.). (2000). *The necessity of informal learning*. The Policy Press.

Confrey, J. (2008, July). A synthesis of the research on rational number reasoning: a learning progressions approach to synthesis. *Paper presented at the 11th International Congress of Mathematics Instruction*, Monterrey, Mexico.

Dahsah, C., & Coll, R. K. (2007). Thai grade 10 and 11 students' conceptual understanding and problem-solving ability in stoichiometry. *Research in Science and Technology Education, 25*, 227–241.

Erceg, N., Aviani, I., Mešić, V., Glunčić, M., & Žauhar, G. (2016). Development of the kinetic molecular theory of gases concept inventory: Preliminary results on university students' misconceptions. *Physical Review Physics Education Research, 10*(2), 020139.

Fang, S. C., Hart, C., & Clarke, D. (2014). Unpacking the meaning of the mole concept for secondary school teachers and students. *Journal of Chemical Education, 91*(3), 351–356.

Fielding-Wells, J. (2013). Inquiry-Based Argumentation in Primary Mathematics: Reflecting on Evidence. *Mathematics Education Research Group of Australasia.*

Furio, C., Azcona, R., & Guisasola, J. (2002). The learning and teaching of the concepts 'amount of substance'and 'mole': A review of the literature. *Chemistry Education Research and Practice, 3*(3), 277–292.

Indriyanti, N. Y., & Barke, H. D. (2017, August). Teaching the mole concept with sub-micro level: Do the students perform better?. In *AIP Conference Proceedings* (Vol. 1868, No. 1, p. 030002). AIP Publishing.

Johnstone, A. H. (1971). Topic difficulties in chemistry. *Education in Chemistry, 8*(6), 210–213.

Mansoor, N. I. A. Z., & Rodrigues, M. A. (2001). Do we have to introduce history and philosophy of science or is it already 'inside'chemistry?. *Chemistry Education Research and Practice, 2*(2), 159–164.

Milton, M. J. (2013). The mole, amount of substance, and primary methods. *Metrologia, 50*(2), 158.

Miyakawa, T., & Winsløw, C. (2009). Didactical designs for students' proportional reasoning: An "open approach" lesson and a "fundamental situation". *Educational Studies in Mathematics, 72*(2), 199–218.

Musa, U. (2009). Teaching the mole concept using a conceptual change method at the college level. *Education, 109*(4), 683–691.

Niaz, M. (1989). The role of cognitive style and its influence on proportional reasoning. *Journal of Research in Science Teaching, 26*(3), 221–235.

Scott, F. J. (2010). Is mathematics to blame? An investigation into high school students' difficulty in performing calculations in chemistry. *Chemistry Education Research and Practice, 13*(3), 330–336.

Shadish, W. R., Cook, T. D., & Campbell, D. T. (2002). *Experimental and quasi-experimental designs for generalized causal inference*. Houghton Mifflin.

Siswaningsih, W., Firman, H., & Khoirunnisa, A. (2017, February). Development of two-tier diagnostic test pictorial-based for identifying high school students misconceptions on the mole concept. In *Journal of Physics: Conference Series* (Vol. 810, No. 1, p. 010117). IOP Publishing.

Sokolowski, A. (2019). Modeling the mole understanding with mathematical reasoning. *International Journal of Physics & Chemistry Education, 11*(4), 85–92.

Staver, J. R., & Lumpe, A. T. (1995). Two investigations of students' understanding of the mole concept and its use in problem-solving. *Journal of Research in Science Teaching, 32*(2), 177–193.

Tullberg, A., Strömdahl, H., & Lybeck, L. (1994). Students' conceptions of 1 mol and educators' conceptions of how they teach 'the mole'. *International Journal of Science Education, 16*(2), 145–156.

Uce, M. (2009). Teaching the mole concept using a conceptual change method at the college level. *Education, 109*(4), 683–692.

Chapter 11
Enabling Covariational Reasoning in Einstein's Formula for Photoelectric Effect

11.1 Prior Research

The research identified several students' misconceptions regarding the photoelectric effect's (PE) principles and difficulties with graphs' interpretations. McKagan et al. (2009) found that 42% of students "claimed (mistakenly) that a voltage is necessary and sufficient for current flow or to overcome the work function of the metal" (p. 92). Klassen (2008, July) found out that the main difficulty and, thus, the source of misconceptions originates from incorrect interpretation of the work function. Steinberg et al. (1996) reported that students faced problems in making predictions of relations embedded in the photoelectric effect. A series of experiments, computer-based tutorials, and simulations have been developed to help students understand the photoelectric effect. Earle et al. (2003) designed an experiment to help students understand that a photocurrent is dependent on *work function* when photons were incident on one side of a thin gold film, and electrons were ejected. Calvin (2004) created a module on quantum physics that had students reproduce some of the reasoning that won Einstein his Nobel Prize. Kovačević and Djordjevich (2006) developed a mechanical model of the photoelectric effect that enabled instructors to apply an analogy to support its understanding. Steinberg et al. (1996) created a computer-based tutorial on the photoelectric effect that addressed physics students' conceptual and reasoning difficulties. Sokolowski (2013) proposed an inductively designed instructional unit highlighting the PE's main phases and their entanglements. Research conducted by McKagan et al. (2009) proved that the virtual lab that emphasized the underlying principles of the photoelectric effect "provided significant improvement over traditional instruction," even though as they claimed, "there is still room for improvement in developing students' skills in reasoning from observations to inferences" (p. 94). While all these efforts aim to develop a conceptual understanding of the PE, the algebraic representation of the law of conservation of energy is not being analyzed for its adherence to algebraic function rules. Research

A. Sokolowski, *Understanding Physics Using Mathematical Reasoning*,
https://doi.org/10.1007/978-3-030-80205-9_11

shows that students consider functions as conceptually richer mathematical structures to learn about phenomena behavior than formulas (Sokolowski, 2020a); therefore, attempts to exemplify formulas' covariational attributes will be a priority in this study. While a linear function is widely taught and applied in the high math school curriculum, its interpretation in PE confuses the students. Considering the prior research findings, it was hypothesized that students' difficulties with applying mathematical reasoning to analyze PE might be rooted in how the central equation that models the law of conservation of energy of the PE, $K_{Max} = hf - W_o$, is presented and graphed. The following sections attempt to uncover the potential weaknesses of the equation and suggest its reimaging.

11.2 Theoretical Background

Covariational reasoning was discussed in more detail in Chap. 5. In this section, it is being used as a cognitive activity that involves the coordination of two varying parameters while trying to determine how one parameter affects the other in the context of the photoelectric effect. Covariational reasoning is primarily developed in mathematics classes; however, mathematics as a subject where students learn concepts and ideas without contexts does not offer students many opportunities to apply this type of synthesis in realistic settings. A preliminary survey conducted using academic research engines such as Google Scholar revealed that research in physics on covariational reasoning is staged on the conceptual level. Converting information to its graphical representation as a function requires identifying parameters that change (variables) and their classification as independent and dependent. Physics is a study of continuing dynamic phenomena; thus, it is reasonable to claim that covariational reasoning, as a way of inquiry, should earn its prominent role in physics education. This study is an example of such an undertaking.

11.3 Embracing the PE into the Framework of Covariational Representation

Einstein's formula to quantify the energy in the photoelectric effect includes—directly and indirectly—multiple concepts and principles; the Compton effect, Millikan's experiment, duality of light, photon, quantum, electric current, the intensity of the electric current, work function, threshold frequency, binding energy, potential difference, stopping potential and kinetic energy. All these principles and parameters aid the analysis as they all create a detailed picture of the phenomena' behavior. Following this complexity, a diversity of graphical representations is developed to support its quantitative and qualitative understanding. In this paper, a scrutiny of the main formula to quantify the energy in the photoelectric effect, $K_{Max} = hf - W_o$, will result in its reasembling to a covariate expression.

While the students' difficulties have a wide range, a substantial focus is given to identifying students' confusion with interpreting the graph of $K_{Max} = hf - W_o$. Following this finding, a discussion will be initiated to address this challenge by analyzing the graph of K_{Max}. This analysis will reflect on the effects of addends in the PE formula in exemplifying a linear covariation between external photons' frequency and the ejected electrons' kinetic energy. As a result, a piecewise linear function will emerge and be proposed in Sect. 13.4.

11.3.1 Weaknesses of the Graph of K_{MAX} Versus Photons' Frequency Presented in Physics Resources

The paper's purpose was to provide alternative graphical and algebraic representations of the PE. It was hypothesized that students are confused with the graph interpretation because the currently used algebraic and graphic models of PE do not foster a mutual coherence with its scientific understanding. The following sections provide support for that claim. The scrutiny will begin from analysing the graphical representation of the PE formula, shown in Fig. 11.1, and it will proceed toward discussing its algebraic equivalent. As the discussion unfolds, suggestions for modifications will emerge.

Figure 11.1 illustrates a graphical representation of $K_{Max} = hf - W_o$ that can be found in physics textbooks ranging from high school to college levels, e.g., Servay and Faughn (2002, p. 835), Allum and Talbot (2018, p. 509), Cutnell and Johnson (2007, p.119), and Giancoli (2005, p.760). The graph (Fig. 11.1) is meant to represent a linear function $f(x) = mx - b$. Some graph elements might challenge the students' mathematical reasoning if the general slope-intercept function is mapped into the PE formula. At first, the dotted part of the graph suggesting the electrons' zeroth kinetic energy can be misleading. In mathematics, a dotted line usually depicts attributes like function vertical or horizontal asymptotes. None of such graph attributes is meant to be labeled on this graph. Thus, to denote the electrons' zeroth kinetic energy, a different graphical representation should have been used. There are no frequency restrictions on the energy function $K_{Max} = hf - W_o$; thus, from a mathematical standpoint, there should be a way to evaluate the electrons' kinetic energy for any photons' frequency. This statement challenges the observation of the PE; the kinetic energy of the ejected electrons has a value of zero for $0 \leq f \leq f_o$. Does the graph intend to inform that the electrons' have negative kinetic energy or negative work function on its dotted part? Can the electrons' kinetic energy be concluded when its graph lays below the frequency axis? These are sample questions that students might have while trying to find a physical sense of such a graph using their math knowledge. Problematic can also be the interpretations of the graph's horizontal and vertical intercepts, which can stem from labeling two different quantities on

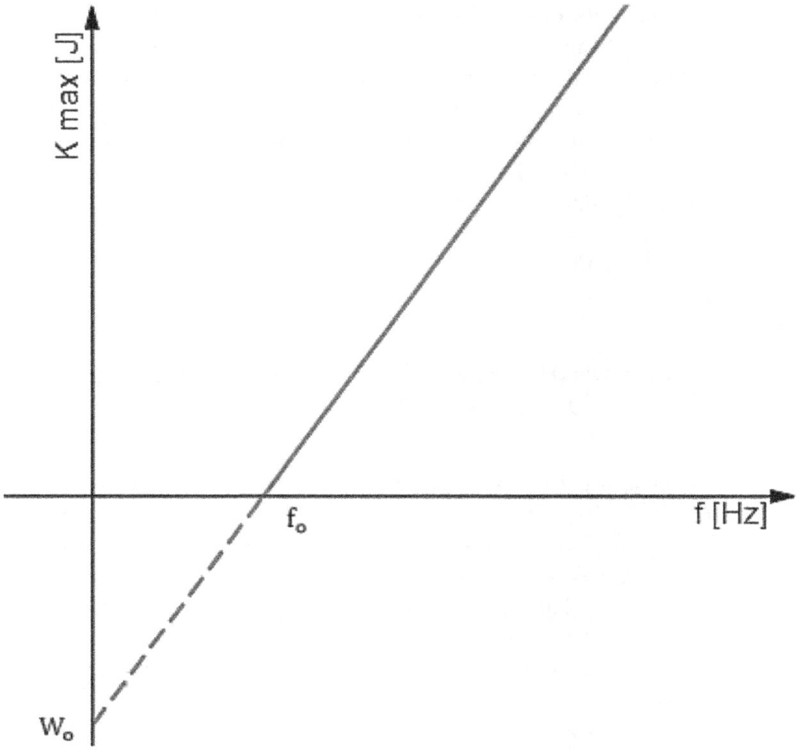

Fig. 11.1 A typical graph of the law of conservation of energy for PE found in physics textbooks

the graph's vertical axis: kinetic energy and the work function. In mathematics, the vertical intercept is a particular value of $f(x)$, often called the initial value, whose units are similar to that of $f(x)$. Work function represents in the equation $K_{Max} = hf - W_o$, *per see* the vertical intercepts of the linear function. However, this interpretation does not entirely reflect the scientific underpinnings of the PE formula. When the photons' frequency $f = 0$, there is no current observed, which is depicted by $K_{Max} = 0$. Thus, the vertical intercept of K_{Max} should be equal to zero. According to the PE formula, if $f = 0$, $K_{Max} = -W_o$. While the scientific interpretation can be (and typically is) augmented to satisfy such graphs, the algebraic reasoning does not fully support the scientific underpinnings if students attempt to interpret it directly using their mathematics knowledge. This element might cause some uncertainties; thus, a need emerged to re-evaluate and assemble the PE formula in a way that generates sound scientific conclusions coherent with observations. Students realize that kinetic energy is non-negative and that the external energy must be supplied to the system to free the electrons. This scientific fact has its right in the theory of algebraic equation; however, its covariational representation calls for revisiting. Suppose that one attempts to find the horizontal intercept of the graph. Students realize that in the search for the x-intercept and its scientific interpretation, they let the dependent parameter, thus $K_{Max} = 0$. Using the formula $hf = W_o$, which corresponds to the phase of freeing the electrons from the metal plate. In this case, $f_o = \frac{Wo}{h}$ is the threshold frequency of the metal plate or the incoming photons'minimum frequency that ejects the atoms' electrons. The kinetic energy of the electrons is still zero at this phase of analysis. Can the threshold frequency be called the x-intercept of the graph of K_{Max}?This is doubtful. According to the rules of algebra, x-intercepts are the x-values for which $f(x) = 0$. The function $f(x)$ that K_{Max} represents should have a value of zero on the entire interval $0 \leq f \leq f_o$. This behaviour is supported by the fact that if $0 \leq f < f_o$ no electrons are ejected or they are ejected with a zero kinetic energy when $f = f_o$. The electrons are ejected when the frequency of photons reaches the value of the threshold frequency f_o. However, they still have no kinetic energy to reach the collector. This scientific fact is not explicitly conveyed to students using $K_{Max} = hf - W_o$. The zeroth kinetic energy of the electrons on $0 \leq f \leq f_o$ will not emerge from using the algebraic form $K_{Max} = hf - W_o$. On the graph, this inquiry leads to the controversial dashed line (Fig. 11.1). If one attends to identifying a relationship between the photons' frequency and the ejected electrons' behavior, then the dashed part calls for a revision. It is hypothesized that students' confusion regarding the differences between the interpretation of the work function and electrons' kinetic energy might stem from trying to find a mathematical-physical sense of this part of the graph. The section that follows discusses the algebraic structure of the law of conservation of energy and attempts to align it with the phenomenon's behaviour.

11.3.2 Covariation of Photon's Energy and Frequency as a Linear Function

Photon's energy is commonly presented as $E = hf$. The scientific interpretation of the formula is straightforward. Photon's energy depends on the light source's frequency, with Planck's constant considered a proportionality constant or the relation's slope. A study (Sokolowski, 2020b) revealed that physics students have difficulties classifying parameters embedded in traditionally expressed physics formulas as dependent and independent, which creates uncertainty with graph interpretations. Thus, when covariational reasoning is concerned, function notation that signifies parameters' unique behaviour within the system is employed, which results in the formula expressed as $E(f) = hf$. This expression has a form of a linear function, and its slope must have a definite interpretation. Slope interpretations that link mathematical and scientific underpinnings also require attention to be correctly introduced to science students (Kwon et al., 2000). The unit of the slope in the PE is Js. The product, though, does not explicitly foster the slope interpretation as a rate. The slope in mathematics is considered as $m = \frac{\Delta y}{\Delta x}$, thus a rate or ratio of two quantities. If applications are concerned, the scientific units resulting from multiplying the slope and the independent variable must yield the unit of the variable of interest or the dependent variable. Does the formula $E(f) = hf$ convey the unit? In $E(f) = hf$, the Planck's constant $h = 6.626 \bullet 10^{-34}$ Js. Planck calculated the value from experimental data on black-body radiation and expressed the constant as a product of two quantities. While the unit's interpretation as a product is not an obstacle for physics instructors, it might not benefit the students. It is suggested that the unit be reconstructed to reflect on the slope interpretation more explicitly yet retain its original scientific sense. Since the unit of *hertz* is labelled on the horizontal axis, Planck's constant might be expressed in the units $\frac{J}{Hz}$, which is mathematically equivalent to *Js*. Expressing the Planck's constant as a rate of the energy change per unit of frequency allows enacting its precise interpretation using mathematical reasoning (Sokolowski, 2018); the energy of a photon changes by $6.626 \bullet 10^{-34} J$ per 1 Hz of photons' frequency change. Such interpretation also supports the quantized nature of the photon's energy. It is suggested that the traditional unit of the constant as *Js* is introduced to students after the PE, as a linear relation, is fully understood.

11.3.3 Electrons' Binding Energy as a Function of Photons Threshold Frequency

In the traditional algebraic setting, the independent variable of $f(x)$ affects the dependent. The vertical intercept of $f(x)$ represents a particular value of the dependent variable when the independent variable takes a value of zero. As earlier discussed, the scientific underpinnings of the formula $K_{Max} = hf - W_o$ do not entirely fit into this mathematical model. The binding energy of the electrons is an external parameter built in the phenomenon as an additional covariate. Its inclusion has its right, which is warranted in the work–energy principle stating that photons do work on electrons. Therefore, electrons are ejected. The work function is considered a metal plate parameter, which can be regarded as an additional object, thus a covariate quantity in the PE. From the covariational reasoning standpoint, its interpretation might be challenging to follow the PE dynamics. Therefore, instead of W_o, its equivalent expression hf_o will emerge and be used in the central Eq. (11.2) to quantify the PE. The symbols hf_o will correlate with the symbols of the function's established parameters. Thereby they also will enhance the math–physics interphase of the phenomena representation. The work function W_o will be included in Eq. (11.4) to support PE explorations and problem-solving.

11.3.4 Maintaining a Minimum Number of Covariational Parameters During the Inquiry

The number of varying parameters was initially reduced to assure clarity of the line of investigation. Physics formulas are traditionally built up as entanglements of two variable quantities. The photoelectric effect is more complicated if viewed from this angle. A high spectrum of contributing principles and quantities in the PE demands a high cognitive load; thus, a high capacity of the learners' working memory. Consequently, the multiple arrays of covariations that stem from the photocurrent production, while not requiring an advanced mathematical apparatus, can unintentionally dilute the PE's core idea. To maintain the number of the covariational parameters to its necessary minimum and yet attend to a full spectrum of the PT understanding, the following restrictions were imposed (a) the plate ejecting electrons; thus, hf_o remained unchanged, and (b) the external battery was removed initially from the consideration. Varying the plate's type certainly enriches the exploratory nature of the phenomena; however, this element does not directly contribute to the phenomena' algebraic representation. Analogically, the idea of the stopping potential was not included at first. While having its right by supporting the PE quantification when introduced at the initial stage, it might inadvertently create an impression that the battery enhances the electrons' movements. These steps were taken to eliminate possible distractions and clear the inquiry path to help students realize that the only condition for initiating the photocurrent was ejections of the electrons from the emitter and having them reach the collector. Lloyd (2020) provided further details supporting the removal of the stopping potential from PE teaching. As a result, the plate remained unchanged throughout the experiment demonstration, and the external battery was set up at zero potential. Thus, the analysis was restricted to Einstein's equation. Both these parameters remained silent and labelled as constants throughout the instructional unit conduct. These steps were taken to eliminate possible effects of other covariate parameters and thus clear the inquiry path to help students realize that the only condition for initiating the photocurrent was ejections of the electrons from the emitter and having them reach the collector.

11.4 Reassembling the PE Formula to Assure a Coherence of Representations

Research showed that physics students need to be guided to reimage formulas into algebraic functions (Sokolowski, 2019); more specifically, they can better identify and analyze casual relationships if the physics formulas resemble mathematical functions denoted using function notation. Function notation explicitly highlights the independent and dependent parameters (variables); thus, the path to focus investigator attention on their mutual entanglement is cleared. Following this finding, instead K_{Max}, $K(f)$ will be used to denote the kinetic energy of the ejected electrons. Using K_{Max} has its scientific validity, and it is suggested as an additional fact that enriches the scientific nature of the PE. The proposed notation $K(f)$ could retain the subscript, $_{Max}$. However, in students' realms, such subscript could evoke the maximum value of the kinetic energy, which in its mathematical sense triggers specific behavior and algorithms for its computations; these actions are not intended.

11.4.1 Graph Constructing

Visual representations are to decrease the demand for working memory to allow focusing on understanding the phenomena behavior. However, the graph is to be free from abstractions and clear to read its scientific underpinnings to serve its purpose. Graphs must also be coherent with algebraic representations of the phenomena' processes. Formulating the algebraic expression of the PE will be based on developing its graphical counterpart first. The virtual lab whose snapshot is presented in Fig. 11.2 enriches the visual representation and serves as a source of scientific validation of the enacted algebraic expression.

The experiment provided evidence that no electrons were ejected when $0 \leq f < f_o$. This scientific fact demanded that the kinetic energy graph should have zero value for that frequency range. Consequently, this observation suggested a piece-wise linear function and that the first piece of the $E(f)$ graph should have a zero slope and be positioned along the frequency axis. When the frequency of the photon reached the threshold value, f_o, electrons were ejected. A further increase in the photons'

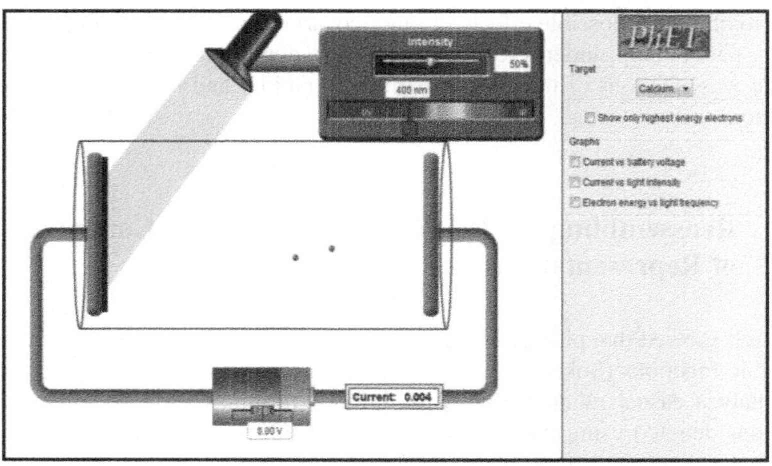

Fig. 11.2 Demonstration of photons kinetic energy. (Source: PhET Interactive Simulations (n.d.))

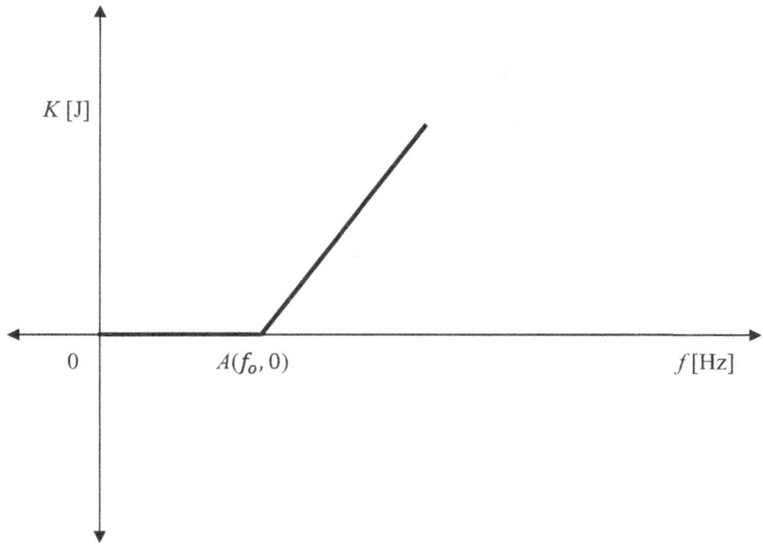

Fig. 11.3 A piecewise linear function representing kinetic energy of the ejected electrons

frequency caused a rise in the electrons' speeds, thus their kinetic energy. The electrons' kinetic energy depended linearly on the incoming photon's frequency. While justifying the linear covariance is difficult by observing the electrons' speeds, this can be followed by activating *Electrosn energy versus light frequency graph* made available on the right side on the simulation menu (Fig. 11.2). Therefore, the second part of the graph must also be linear but with a slope different than zero. A graph representing such a relation is illustrated in Fig. 11.3.

The graph does not cross the horizontal axis because kinetic energy cannot be negative. The function satisfies the vertical line test, following its math conditions (Stewart, 2000). It also meets the principle of continuity in its domain. Thus, mathematically, the graph depicts a linear covariate function that corresponds with the phenomena' behavior. In the earlier study (Sokolowski, 2021), the left and right-sided notations were denoted on the graph to indicate the experiment's possible directions to attain the threshold frequency. However, it seems that such insertion is not necessary.

11.4.2 Finding Algebraic Representation of the Graph

The horizontal and the inclined parts of the graph signify different algebraic rules within the function, yet two mutually consistent scientific behaviors. The vertical axis of the graph represents the kinetic energy of the ejected electrons on its entire range. Since the kinetic energy can be zero or greater than zero, the graph lays along and above the horizontal axis.

With such scientific and algebraic background, the students are ready to find the graph's algebraic representation. If the external frequency of the photons is not sufficient to overcome the electrons' binding energy, no electrons will be ejected, which is algebraically represented by $K(f) = 0$. Thus, according to the graph, the following can be concluded: $K(f) = 0$, when $0 \le f \le f_o$. If the external frequency reached f_o, the electrons are ejected. However, they have no capability of moving toward the collector. This scientific fact could be included as an additional third part of the function. Nevertheless, it was seen as an element of overusing the mathematical description and diminishing its clarity. To find the equation for the second piece of the graph, one can refer to the slope-point form for a linear function $y - y_A = m(x - x_A)$. The slope of the function is the earlier discussed Planck's constant. By assigning the experiment parameters to the variables, one learns that $K(f) - y_A = h(f - x_A)$. To complete the function, at least one additional coordinate from the graph of $E(f)$ is necessary. If the frequency of incoming photons reached the threshold frequency, f_o, $K(f_o) = \lim_{f \to f_o^-} (hf - hf_o) = hf_o - hf_o = 0$. The electrons are ejected, but they do not possess kinetic energy. The additional coordinate necessary to formulate $K(f)$ is then $A(f_o, 0)$. Substituting this coordinate for (x_A, y_A) in the slope-point linear model results in $K(f) - 0 = h(f - f_o)$, which leads to $K(f) = hf - hf_o$ when $f > f_o$. By combining both pieces of the function, one learns that:

$$
\begin{aligned}
&K(f) = 0, \quad \text{when} \quad 0 \le f \le f_o \\
&K(f) = hf - hf_{o,} \ \text{when}\ f > f_o
\end{aligned}
\tag{11.1}
$$

The function $K(f)$ presented in (11.1) is a piecewise linear function, with two different rules on its domain. The parameter f_o represents the metal plate's threshold frequency and a specific value of the frequency of the incoming photons that ejects the electrons. It is to note that $f_{o,}$ called *cut off frequency*, has its specific value on the domain of $K(f)$, not on its range. Considering this condition, the kinetic energy function of the ejected electrons, $K(f)$, can be written in a more concise fasion using a piecewise function representation:

$$
K(f) = \begin{cases} 0, & \text{when}\, 0 \le f \le f_o \\ hf - hf_{o,} & \text{when}\ f > f_o \end{cases}
\tag{11.2}
$$

The electrons are not ejected at all if $f < f_o$. Students are to realize that when $f = f_o$, electrons are ejected but with a zero speed. This fact can also be inferred from the first piece of the function. If the frequency of the incoming photons is greater than that of the threshold frequency, thus if $f > f_o$, then electrons are ejected. According to the law of conservation of energy for the PE, electrons possess kinetic energy that can be computed from $K(f) = hf - hf_o$. The function illustrates the dynamic process of the phenomena, and the mathematical reasoning is aligned with it. Such function also satisfies the principle of continuity on its entire domain, $f \ge 0$, which transcends to the fact that the kinetic energy of the electrons can be calculated for any frequency within that domain. There can be a restriction on frequency maximum value due to the range of the light frequencies available for a specific experiment.

11.4.3 Linking the Photons Threshold Frequency and the Work Function $hf_o = W_o$

While the derived piecewise function satisfies the rules of algebra, students welcome its relevance to problem-solving. Using the law of conservation of energy, the threshold energy of the incoming photons is equal to the work function of the emitter, $hf_o = W_o$.

$$\begin{cases} K(f) = \begin{cases} 0, \text{ when } 0 \le f \le f_o \\ hf - hf_o, \text{ when } f > f_o \end{cases} \\ hf_o = W_o \end{cases} \quad (11.3)$$

Creating such an algebraic system of equations allows linking W_o with $K(f)$, which is accomplished by replacing hf_o in $K(f)$ by W_o. By such algebraic manipulations, $K(f)$ takes the following form:

$$K(f) = \begin{cases} 0, \text{ when } 0 \le f \le f_o \\ hf - W_o, \text{ when } f > f_o \end{cases} \quad (11.4)$$

Once the students realize that the system of equations allows including W_o in the function equation, the confusion about the dashed line (Fig. 11.1) in the general graphical representation of the PE can be eliminated. Using W_o relates all involved entities in the photoelectric effect and supports problem-solving. However, it does not directly support the understanding of the virtue of the photoelectric effect and its graph. Certainly, including W_o, the work function of the emitter extends mathematical reasoning, thus enriches the inquiry. Using various work functions, thus different objects (emitters) will generate different threshold frequencies and strech or compress the graphs horizontally. However, the slopes of all graphs will remain identical and equal to Planck's constant.

11.5 Summary and Conclusions

The form of the law of conservation of energy for PE found in the textbooks works fine to support problem–solving; however, its graphical representation as constructed directly using its algebraic formula is problematic. The paper's purpose was to indicate its weaknesses as seen from the students' perspective and propose its alternative algebraic representation coherent with the phenomena' scientific underpinnings and a more rigorous algebraic notation.

While a formal evaluation of the proposed algebraic representation of the photoelectric effect was not possible to furnish at the time of the study development, an instructional unit with its context was delivered to a group of twenty (N = 20) high school physics students. The unit was conducted over two class periods. On the first day, the students investigated the scientific underpinnings of photons' energy and interactions with different metals, $W(f_o) = hf_o$ that generated a linear function with the slope of h. During the second day, PE's virtual setup was used, and the mathematical modeling of the PE proceeded. All students were familiar with formulating and interpreting piecewise functions. Thus, the process of enacting the graphical and algebraic representations moved smoothly. Graph (Fig. 11.1) was not discussed during that time. It was discussed on another occasion when the students learned the purpose of inserting the external battery into the circuit. While quantum physics is considered a complex physics section, the students appreciated the lesson's simplicity and the dynamic simulated processes that provided them with model-like scenarios and a better understanding. The lab helped the students construct the mental images of PE that enhanced the functional covariations of the parameters of interest. The students were excited to see applications of piecewise functions in science. While they have explored these functions' visual representations in kinematics and energy sections, piecewise formulas are rare in quantum physics. As a follow-up task, they were assigned to construct piecewise functions for the given threshold frequency of selected metal plates, e.g., sodium and gold. It was interesting to realize that knowing this parameter was sufficient to set up boundaries to precisely represent the PE's behavior graphically and algebraically.

Students' genuine involvement and participation illustrated that mixing mathematical structural knowledge with scientific inquiry generates productive learning environments worthy of further explorations. Emphasizing the coherence of scientific evidence and consistency of algebraic representations was also to send a message that although physics is considered a mature academic discipline, precision of algebraic embodiments of its laws can support its better understanding. Thus, an opportunity to formulate new models and representations of natural phenomena exists.

References

Allum, J., & Talbot, C. (2018). *Physics for the IB diploma, hodder education* (p. 509).

Calvin, S. (2004). Following in Einstein's footsteps: Teaching the photoelectric effect. *The Physics Teacher, 42*(6), 340–341.

Cutnell, J., & Johnson, K. W. (2007). *Physics* (Vol. 2, 7th ed., p. 919). New Wiley & Sons, Inc..

Earle, G. D., Copp, B. L., Klenzing, J. H., & Bishop, R. L. (2003). A novel empirical study of the photoelectric effect in thin gold films. *American Journal of Physics, 71*(8), 766–769.

Giancoli, D. C. (2005). *Physics: Principles with applications* (6th ed., p. 760). Prentice-Hall.

Klassen, S. (2008, July). The photoelectric effect: Rehabilitating the story for the physics classroom. In *Proceedings of the Second International Conference on Story in Science Teaching*, pp. 1–17.

Kovačević, M. S., & Djordjevich, A. (2006). A mechanical analogy for the photoelectric effect. *Physics Education, 41*(6), 551.

Kwon, Y. J., Lawson, A. E., Chung, W. H., & Kim, Y. S. (2000). Effect on development of proportional reasoning skill of physical experience and cognitive abilities associated with prefrontal lobe activity. *Journal of Research in Science Teaching, 37*(10), 1171–1181.

Lloyd, D. R. (2020). Should the concept of stopping potential be removed from teaching for A-level? *Physics Education, 56*(1), 015501.

McKagan, S. B., Handley, W., Perkins, K. K., & Wieman, C. E. (2009). A research-based curriculum for teaching the photoelectric effect. *American Journal of Physics, 77*(1), 87–94.

PhET Interactive Simulations. (n.d.). The University of Colorado at Boulder. Retrieved from http://phet.colorado.edu, March 2021.

Servay, R., & Faughn, J. (2002). Holt Physics, Holt, Rinehart and Winston, Austin, USA, p. 835.

Sokolowski, A. (2013). Teaching the photoelectric effect inductively. *Physics Education, 48*(1), 35.

Sokolowski, A. (2019, August). Enhancing scientific inquiry by mathematical reasoning: the case of applying limits to model motion of a system of objects. In *Journal of Physics: Conference Series* (Vol. 1287, No. 1, p. 012051). IOP Publishing.

Sokolowski, A. (2020a). Unpacking structural domain of mathematics to aid inquiry in physics: A pilot study. *Physics Education, 56*(1), 015009.

Sokolowski, A. (2020b). Developing Covariational reasoning among students using contexts of formulas: (are current formula notations in physics aiding graph sketching?). *The Physics Educator, 2*(04), 2050016.

Sokolowski, A. (2021). Enabling covariational reasoning in Einstein's formula for photoelectric effect. *Physics Education, 56*(3), 035029.

Steinberg, R. N., Oberem, G. E., & McDermott, L. C. (1996). Development of a computer-based tutorial on the photoelectric effect. *American Journal of Physics, 64*(11), 1370–1379.

Stewart, J. (2000). *Calculus concepts and contexts* (2nd ed., p. 45). Brooks/Cole.

Chapter 12
Are Physics Formulas Aiding Covariational Reasoning? *Students' Perspective*

12.1 Introduction and Prior Research Findings

Students acquire graphing skills in mathematics by developing covariational reasoning (Carlson et al., 2002). This competence is rooted in identifying independent and dependent variables in the given context and converting it into graphs satisfying the context's attributes and conditions for being functions. Physics students use quantities, and at large, formulas in their physics courses to solve problems and to graph relationships between quantities. Contemporary research on understanding physics formulas is often motivated by how physics students interpret these expressions in the physical sense (e.g., Domert et al., 2012). Little research is dedicated to how students perceive physics formulas in the mathematical sense. Focusing on formulas' physical sense has its own right, and it develops the notion that physicists are focusing on interpretations of symbols in formulas using different semiotics (Redish, 2006). Such research generates findings that support that aim; Sherin (2001) found out that students attempt to understand physics formulas by the vocabulary of formulas' component symbols. University physics students display their understanding of physics formulas by interpreting the symbols that formulas are built from (Domert et al., 2012). Dunn and Barbanel (2000) and McDermott et al. (1987) concluded that students' weak graphing skills might not necessarily originate from their inadequate math skills. In a similar accord, Hammer and Elby (2003) and Ceuppens et al. (2019) argued that students possess the needed math skills and knowledge but fail to activate these assets in physics. Sokolowski (2017, 2019) found out that physics students are not sensitive to verifying graphs for adherence to principles of functions, especially when continuity and differentiability are concerned. This finding ultimately led to the conclusion that physics graphs do not necessarily stem from covariational relationships from students' standpoints, even though they are supposed to depict functions in their mathematical sense.

A. Sokolowski, *Understanding Physics Using Mathematical Reasoning*, https://doi.org/10.1007/978-3-030-80205-9_12

Physics contexts often surface in mathematics assessment items and research. A handful of these findings revealed that students have difficulties reading and interpreting graphs of motion in the context of reality (Nemirovsky & Monk, 2000). Students face problems distinguishing between a position-time graph and the actual path of the object's motion (Leinhardt et al., 1990). Some students confuse the object's velocity with its position (Monk, 1992). Other studies disclosed difficulties with sketching velocity and acceleration–time graphs provided with a position-time graph, which, according to Harel and Confrey (1994), is due to a weak understanding of the concept of rate. Bonotto (2013) showed that even when the problems and methods encountered in the classroom were similar to the real situations, students tended to disassociate the analysis from the problem-solving techniques learned in the school. Dufresne et al. (1997) found that problem-solving in introductory mathematics and science courses is often algorithmic or schematic that turns students to mimicking the teachers or textbook procedures. In an attempt to find a reasonable explanation for such difficulties, Planinic et al. (2013) stated that these difficulties are due to the nature of the contextual problem that requires interpretation and translation of meanings into mathematical language.

Identifying the sources of potential weaknesses in students' understanding is one step toward improvement; suggesting instructional interventions for that improvement is a way to make things better. Domert et al. (2012) formulated the epistemological *mindsets* of themes to make students aware of the importance of formulas in physics learning. These mindsets offered students practice recognizing symbols in formulas and opportunities to learn how and when to use these formulas. Cui et al. (2006 February) turned to mathematics and advocated for including more physics applications in mathematics courses. Sokolowski and Capraro (2013) focused on the differentiation between the position graph and the object's path of motion. They developed a modeling activity that offered students opportunities to contrast both these representations in considering their functional attributes. Uhden et al. (2012) proposed a scheme where mathematization and interpretation of physics formulas were grounded in their conceptual predicaments. More recently, Zimmerman et al. (2019) found out that students in introductory physics courses are not usually exposed to physics-specific reasoning skills in their math classes and suggested developing teaching strategies reflecting that reasoning.

Despite physics and mathematics research communities' efforts, students struggle to graph physics contextual problems in mathematics and apply the functions' attributes to display graphically such relationships in physics. The preliminary synthesis of research illustrates a strong emphasis on physical interpretations of symbols used in formulas (e.g., Domert et al., 2012; Uhden et al., 2012), which seem to overshadow the potential of unfolding covariational entanglement stored in the arrangements of the symbols. Focusing on the interpretation of a specific formula might not generalize these structures' properties as a means of cognitive prompts to bring forth their potential for sketching. It is prudent that formulas in physics are to support the quantification of phenomena or their contextual aspects. Though they are still founded on algebra rules, their covariational relationships seem to be missing from being conveyed to students. This paper is to shed more light on algebraic interpretations of formulas as sketchable entities. Departing from this perspective, an attempt will be made to develop a broader reasoning spectrum on formulas' interpretation in their sketching. In its initial form, this study was published in *The Physics Educator,* (Sokolowski 2020b).

12.2 Theoretical Background and Methods

12.2.1 Foundations of Covariation Reasoning

Detailed underpinnings of covariational reasoning were presented in Chap. 5; therefore, the following section is a general introduction to that subject.

Studies show that the function's perception as a process that accepts specific inputs and produces outputs is essential for a mature image of function (Johnson, 2015). This entanglement is advanced in mathematics courses by covariational reasoning. The process begins with identifying the parameters and coordinating their intermediate relationships. Following this process, images of covariation are considered developmental phases that conclude with the formulation of the entire event under consideration. Research in mathematics education (Thompson et al., 2017) suggests that curriculum and instruction should increase emphasis on advancing math students from a coordinated image of two variables to a coordinated image of their instantaneous rates of change. Such analyses would allow for zooming into the dynamics of the mathematized phenomena using designated quantities. Covariational reasoning is being researched in physics education as well, yet the impact on teaching physics is not satisfactory. One of the essential mental actions to acquire fluency in covariational reasoning is coordinating the value of one variable with changes in the other (Carlson et al., 2002). Can omitting strict mathematical notations and parameter classifications in physics lead students to effectively using their math reasoning skills about graphing in physics? The goal of this research is to assess students' abilities in this domain. Reforms in mathematics education (Hughes-Hallett, 2006) emphasize presenting functions in various forms; graphical, symbolic, verbal, and numerical to meet the demands of students' multiple learning styles and train students to make transitioning between these representations. Such didactical enterprises also inspired this study.

12.2.2 Study Description, Participants, Research Questions, and Evaluation Instrument

This study can be classified as diagnostic research following the description by Ketterlin-Geller and Yovanoff (2009). Diagnosis is an integral part of instructional decision-making in education. As the bridge between identifying students' weaknesses and delivering designed interventions, diagnosis also provides valuable information about students' misconceptions in the targeted domain (Fuchs et al., 2003). Diagnostic research is mainly applied for: cognitive assessments, skills analysis, or error analysis (Leighton & Gierl, 2007). In this study, the diagnostic test was used to assess students' skills. More specifically, a diagnosis of students' interpretation of selected mathematics structures used in physics will be scrutinized. Students' responses will be debriefed to shed light on possible interventions to strengthen the flow of students' covariational reasoning from mathematics to physics.

The study attempted to address the following questions:

1. *To what extent can the algebraic structure of a formula affect activating its covariational features related to sketching?*
2. *To what extent can physics students correctly classify mathematical symbols in formulas based on provided real context?*

The study participants consisted of 30 first-year college physics students (18 males and 12 females). Twenty-five of these students (N = 25, 83%) took one or two high school physics courses, and five (N = 5, 17%) took this course physics course for the first time. All students possessed a calculus background. The assessment instrument consisted of a kinematics context followed by two questions.

Starting from a position denoted as x_1, *an object moves with a constant velocity,v, along a horizontal line. Time t is expressed in seconds and the position in meters. Two symbolic representations are depicting the object's position:* $x(t) = vt + x_1$ *and* $x = vt + x_1$.

(a) *Classify each symbol in each equation as a parameter, independent variable, dependent variable, or none. Support your judgment.*
(b) *Can each of these equations* $x(t) = vt + x_1$ *and* $x = vt + x_1$ *depict the motion in x versus time axes graphically?*

Students took the assessment test during the second day of instruction. Thus, one can assume that most answered these questions, referring to their prior high school experiences. The first question was to diagnose students' ability to transform the physics equations into functions and categorize their symbols according to their algebraic interpretations. The algebraic forms of both formulas were identical, the difference laid in denoting their right sides; the first; $x(t) = vt + x_1$ was expressed in the function notation, the other $x = vt + x_1$ in the formula notation. Presenting these equations *side by side* was arranged on purpose. It was assumed that students who would be biased by anticipating differences would not apply covariational reasoning to the extent that would eliminate the bias. Though, students were provided with a physical context that was modeled by these equations.

12.3 Data Analysis

Data analysis consisted of two parts; a descriptive evaluation of the responses presented in Table 12.1 and debriefing analyses of responses whose samples are clustered in Tables 12.2 and 12.3.

Descriptive analysis revealed that the responses could be classified using four categories depending on the individual perception. These categories are described in Table 12.1. The number of students belonging to each category is expressed in a percent form.

Classifying the symbols for position expressed in the function notation appeared more understandable and accessible to extract the physics than that shown in the formula notation. Students are acquainted with $f(x)$ interpretation as f depending on x, or as x being the independent variable of f. This familiar notation steered them to classify t, time as the independent variable and, $x(t)$ the dependent one represented by the object's position. By identifying $x(t)$ as a function, 93% (N = 28) of these students concluded that this formula could be sketched. Thus, one can conclude that they realized a functional covariation of t and $x(t)$. The remaining symbols v and x_1 were classified as parameters by 90% for these students. The classification turned in a different direction if the formula notation $x = vt + x_1$ of the same expression - and representing the same context - was considered. Without highlighting t in the function notation as the independent variable, 13% (N = 4) considered v and x_1 the parameters and the same low percentage of students claimed that the equation could be sketched.

Similarly, a small percentage of students classified t and x as independent and dependent variables. Most of these students (N = 26, 85%) considered the symbols x_1, v, and t as all parameters or independent variables when analyzing the formula notation. These students were asked to support their answers verbally, and sixteen (N = 16, 53%) did so. Table 12.2 provides samples of students' responses solicited verbatim that showed the level of their ability to apply functional notation in physics.

Table 12.1 Descriptive analysis

Equation	Category 1 v and x_1 considered as the parameters	Category 2 t considered the independent variable	Category 3 $x(t)$ or x as the dependent variable	Category 4 The equation can be graphed as given
Functional notation $x(t) = vt + x_1$	85% (N = 26)	90% (N = 27)	90% (N = 27)	93% (N = 28)
Formula notation $x = vt + x_1$	13% (N = 4)	20% (N = 6)	27% (N = 8)	13% (N = 4)

Table 12.2 Students' comments on decomposing $x(t) = vt + x_1$

Student	Response
1	X(t) is a dependent variable, dependent on time, t, which is an independent variable because t is unchanged by other values in the equation but does change on its own rule. V and x_1 are parameters that limit the outcome of the dependent variable, yet they are fixed values.
2	X(t) implies that t is the variable, and the parameter is being evaluated at time t (assuming that t is referring to time). The x1 value is referring to the position of the object, which is independent of how much time has passed, making x_1 a constant value.
3	x_1 furthermore, v are parameters because they are observable values, t is independent because x(t) will vary as t changes.
4	X is a function of t; thus, it is the dependent variable, and t is the independent variable; it is an independent change. V is the parameter as it affects the relationship between x and t. X is constant.
5	X_1 depends on the initial position, meaning it depends on the application, and it is constant within the application. Since it is x(t), this means that x(t) depends on the value of t, not v, so t changes.

Table 12.3 Students' comments on decomposing $x = vt + x_1$

Student	Response
1	X is not a function of t, so each has set values that do not change, so they are all parameters.
2	X is dependent on the three parameters, v, t, and x.
3	V, t, x_1 are all observed values; thus, they all are parameters.
4	X is not defined to be a function of anything so that you could be given the velocity and the starting position. With that, you could find the time that has passed to reach that velocity or any other combination of these variables.
5	All of the letters v, t, and x_1 are not dependent on each other. Thus, they all are independent variables.

Table 12.3 provides verbatim responses that disclose students' attempts to make mathematical sense of the standard formula notation. The purpose of classifying these responses into two tables (categories) was to contrast the students' reasoning that motivated their choices and look for possible factors that deterred them from being correct.

Evaluation of these responses was problematic to a certain degree. These students concluded their answers referring to the interpretation of the functional notation; $f(x)$ or $y(x)$ learned in their mathematics courses. Indeed, without an explicit background, a statement like $z = k + y$ could be classified as built from variables or parameters. Considering such an assumption, some of the answers from Table 12.3 would be marked as correct. However, due to a provided context to be used to support the classification, these answers were not considered correct.

It seems that the absence of function notation in symbolically denoting the object's position confused the students. The majority of these students considered the task associated with $x = vt + x_1$ an example of a plugin and evaluate the formula. It seems that the provided prompts in the context were not attractive enough to suggest a covariational relation. This supports a notion also voiced in other research (e.g., Sherin, 2001; Uhden et al., 2012) that students consider formulas to calculate a specific quantity. A corollary conclusion can be drawn that these science students did not have an opportunity to experience situations when they would be required to envision covariational relations based on the provided context. These conclusions suggest that embracing formulas in functional notation makes the transfer of students' covariational reasoning to physics plausible. An absence of a variable in the function notation confused the identification of the independent variable. It generated - in students' realities - an algebraic structure that is not a function, and thus it is not sketchable. It is to note that the formula selected for the diagnosis has traditionally been used in physics and mathematics courses for problem-solving or sketching purposes. One can claim that if the formula is not expressed in the function notation, students doubt how to find a covariational sense in its algebraic entanglement. They also seem to be lost in assigning specific meanings to the symbols based on their math knowledge, thereby classifying the entire expression as a representation that can be sketched.

There was a visible disconnect between the provided context and the interpretations of the equations while analyzing $x = vt + x_1$. When asked for algebraic interpretations of the symbols, students did not refer to a provided real scenario but referred only to the expression structure. Strong relation to symbolic notations used in mathematics cannot be denied and blamed for students being misled. It should instead be appreciated and taken advantage of to benefit the learning of physics. The reason for that can be a lack of such experiences and heavy emphasis on formulas. This possible mediator could be verified by adding strictly qualitative research to this study and interviewing them, which was unfortunately impossible to arrange. The absence of linking the context with its algebraic embodiment could also be assigned to the responses that did not identify v as the equation's parameter.

Analyzing the structure of formulas and identifying parameters and variables is one element that seemingly calls for more attention; however, analyzing the structures in the context of the given problem for its covariational reasoning is another.

12.4 Summary and Conclusions

This study suggests possible arrays of assisting students to become versed and comfortable in graphing context-driven functional relations in physics. It is also hoped that this small facet has the potential to attract more research and encourage considering more often functional attributes of formulas during physics instruction. This study has certain limitations, one of which is a relatively low sample size. I hope the smaller sample size does not diminish the study's findings as diagnostic research. The larger sample size would undoubtedly increase the validity and allow employing more sophisticated inferential statistics. For example, pairing graphing choices with students' prior math background would shed light on possible mediators. The survey can be easily duplicated in any class setting, conclusions can be drawn, and treatment applied. Another fact that limited the study findings is the uniform mathematical background of the study participants manifested by similar math contents learned following the similar math curricula. Extending the study would shed more light on the effects of different math curricula.

12.4.1 Traditional Formula Notation Does Not Aid Covariational Reasoning in Physics

Formulas are essential to solving problems in science, business, and other academic branches that use quantitative measures. It is understood that function notation is not being used explicitly in these structures. In multivariable expressions, the variable of interest can be represented by any of the component quantities. Moreover, some of the well-known formulas in physics are typically expressed in more than one standard form—for example, $F = ma$ and $a = \frac{F}{m}$. Students might find these algebraic structures confusing when sketching is concerned. Therefore, it is the teacher's task to explain the purpose of these variations following the specificity of the investigation in the plan. It is believed that teaching students how to perceive formulas as covariational entities based on the provided context is essential. This skill can enable them to consider formulas as dynamic functions that can be sketched, differentiated, or integrated. Students need to transfer the reality of an experiment into algebraic symbols and formula equations and analyze the equations as a means of solving for a parameter and deploy further analysis. Envisioning formulas as covariational structures could be enhanced by developing labs where students make connections between what they observe and the mathematical structures they produce. Implementing, suggested in Chap. 5 classification of the parameters, would certainly generalize lab design and its analysis. Emphasizing verification of the coherence between these representations would extend the notion to problem-solving. It is anticipated that such merging would make the students realize that the way mathematical symbols are used in physics must be consistent with mathematics rules. Furthermore, the symbols' interpretations must be compatible with how they were assigned in the lab conducts.

12.4.2 Physics Depends on the Mathematical Rules and Notation

Attempts made in physics courses to improve the flow of math knowledge usually originate from physics points of view, and supportive suitability to algebra rules is established. This flow of knowledge seems to enhance the message that mathematics depends on physics. While, in many cases, this might be true, the literature suggests otherwise. Dirac (1940) once said, "The mathematician plays a game in which he invents the rules while the physicist plays a game in which Nature provides the rules, but as time goes on it becomes increasingly evident that the rules which the mathematician finds interesting are the same as those which Nature has chosen p.124". Understanding physics does depend on understanding and interpreting the rules of mathematics because to numerically express quantity, the rules of mathematics are needed. While physicists use algebraic structures differently in a traditional view because they are interested in analyzing the structures, paying attention to the effects of the precision of mathematics embedded in scientific inquiry design can be equally rewarding.

Referenced earlier, Hughes-Hallett (2006) emphasized presenting functions in various representations to meet the demands of multiple learning styles and train students to transition between these representations. It seems that exercising such transitioning in physics would benefit the learning and understanding of physics as well.

References

Bonotto, C. (2013). Artifacts as sources for problem-posing activities. *Educational Studies in Mathematics, 83*(1), 37–55.

Carlson, M., Jacobs, S., Coe, E., Larsen, S., & Hsu, E. (2002). Applying covariational reasoning while modeling dynamic events: A framework and a study. *Journal for Research in Mathematics Education, 33*(5), 352–378.

Ceuppens, S., Bollen, L., Deprez, J., Dehaene, W., & De Cock, M. (2019). 9th-grade students' understanding and strategies when solving $x(t)$ problems in 1D kinematics and $y(x)$ problems in mathematics. *Physical Review Physics Education Research, 15*(1), 010101.

Cui, L., Rebello, N. S., & Bennett, A. G. (2006, February). College students transfer from calculus to physics. In *AIP Conference Proceedings* (Vol. 818, No. 1, pp. 37–40). AIP.

Dirac, P. A. (1940). XI.—The relation between mathematics and physics. *Proceedings of the Royal Society of Edinburgh, 59*, 122–129.

Domert, D., Airey, J., Linder, C., & Kung, R. L. (2012). An exploration of university physics students' epistemological mindsets towards the understanding of physics equations. *Nordic Studies in Science Education, 3*(1), 15–28.

Dufresne, R. J., Gerace, W. J., & Leonard, W. J. (1997). Solving physics problems with multiple representations. *The Physics Teacher, 35*(5), 270–275.

Dunn, J. W., & Barbanel, J. (2000). One model for an integrated math/physics course focusing on electricity and magnetism and related calculus topics. *American Journal of Physics, 68*(8), 749–757.

Fuchs, L. S., Fuchs, D., Hosp, M. K., & Hamlett, C. L. (2003). The potential for diagnostic analysis within curriculum-based measurement. *Assessment for Effective Intervention, 28*(3&4), 13–22.

Hammer, D., & Elby, A. (2003). Tapping epistemological resources for learning physics. *The Journal of the Learning Sciences, 12*(1), 53–90.

Harel, G., & Confrey, J. (Eds.). (1994). *Development of multiplicative reasoning in the learning of mathematics*. The Sunny Press.

Hughes-Hallett, D. (2006). What have we learned from calculus reform? The road to conceptual understanding. In N. Hastings (Ed.), *Rethinking the courses below Calculus*. Mathematics Association of America.

Johnson, H. L. (2015). Secondary students' quantification of ratio and rate: A framework for reasoning about change in covarying quantities. *Mathematical Thinking and Learning, 17*(1), 64–90.

Ketterlin-Geller, L. R., & Yovanoff, P. (2009). Diagnostic assessments in mathematics to support instructional decision making. *Practical Assessment, Research, and Evaluation, 14*(1), 16.

Leighton, J. P., & Gierl, M. J. (2007). Why cognitive diagnostic assessment? In J. P. Leighton & M. J. Gierl (Eds.), *Cognitive diagnostic assessment for education: Theory and applications* (pp. 3–18). Cambridge University Press.

Leinhardt, G., Zaslavsky, O., & Stein, M. K. (1990). Functions, graphs, and graphing: Tasks, learning, and teaching. *Review of Educational Research, 60*(1), 1–64.

McDermott, L. C., Rosenquist, M. L., & Van Zee, E. H. (1987). Student difficulties in connecting graphs and physics: Examples from kinematics. *American Journal of Physics, 55*(6), 503–513.

Monk, S. (1992). Students' understanding of a function given by a physical model. In G. Harel & E. Dubinsky (Eds.), *The concept of function: Aspects of epistemology and pedagogy*, MAA Notes, 25, 175–193.

Nemirovsky, R., & Monk, S. (2000). 'If you look at it the other way..': An exploration into the nature of symbolizing. In *Symbolizing and communicating in mathematics classrooms: Perspectives on discourse, tools, and instrumental design* (pp. 177–221). Lawrence Erlbaum Associates.

Planinic, M., Ivanjek, L., Susac, A., & Milin-Sipus, Z. (2013). Comparison of university students' understanding of graphs in different contexts. *Physical Review Special Topics-Physics Education Research, 9*(2), 020103.

Redish, E. F. (2006). Problem-solving and the use of math in physics courses. *arXiv preprint physics/0608268*.

Sherin, B. L. (2001). How students understand physics equations. *Cognition and Instruction, 19*(4), 479–541.

Sokolowski, A. (2017). Graphs in kinematics—A need for adherence to principles of algebraic functions. *Physics Education, 52*(6), 065017.

Sokolowski, A. (2019). Graphs in kinematics—A need for adherence to function continuity and differentiability. *Physics Education, 54*(5).

Sokolowski, A. (2020b). Developing Covariational reasoning among students using contexts of formulas: Are the current formula notations in physics aiding graph sketching? *The Physics Educator, 2*(04), 2050016.

Sokolowski, A., & Capraro, M. M. (2013). A constructivist approach to embodying motion problems in mathematics. *Mediterranean Journal for Research in Mathematics Education, 12*(1–2), 121–133.

Thompson, P. W., Hatfield, N. J., Yoon, H., Joshua, S., & Byerley, C. (2017). Covariational reasoning among US and South Korean secondary mathematics teachers. *The Journal of Mathematical Behavior, 48*, 95–111.

Uhden, O., Karam, R., Pietrocola, M., & Pospiech, G. (2012). Modelling mathematical reasoning in physics education. *Science & Education, 21*(4), 485–506.

Zimmerman, C., Olsho, A., Loverude, M., Boudreaux, A., Smith, T., & Brahmia, S. W. (2019). Towards understanding and characterizing expert covariational reasoning in physics. *arXiv preprint arXiv:1911.01598*.

Chapter 13
Adaptivity of Mathematics Representations to Reason Scientifically
Students' Perspective

13.1 Prior Research Findings

Quantitative embodiments of physics laws are rooted in their extensive support by mathematics that provides pathways for setting up the inquiry and serves as a tool to express relationships between investigated quantities in symbolic or graphical forms. While mathematics is used extensively in physics, the knowledge about the supportive use of mathematics in physics education is still being developed (Uhden et al., 2012). The preliminary synthesis of research presented herein pertains to physics students' perception of the tools of mathematics used in physics and improvements suggested by the research. Sherin (2001) found out that physics students learn to understand physics formulas by the vocabulary of component elements rather than their algebraic structures. Domert et al. (2007) found out that even university students' majoring in physics typically do not highlight functional structures of physics formulas and instead focus on interpreting algorithms of operations by rote. Chi (2005) declared that using mathematics tools reduced to operations on formulas reflects students' lack of mathematical and problem-solving skills and a shallow understanding of the underlying mathematics concepts.

Attempts to improve students' transfer of the structural domain of mathematics knowledge to physics beyond the skills of manipulating formulas are undertaken. Domert et al. (2007) proposed the epistemological *mindsets* of themes to make students aware of the significance of formulas in physics. These mindsets included recognizing symbols and algebraic structures in formulas and how and when to use these formulas in everyday life. Consequently, a discussion of the meanings of the formulas was encouraged. Hestenes (2003, p. 104) called for more research targeting the design and use of mathematics in physics education. Cui et al. (2006, February) reflected on calculus and advocated for including more physics applications to prepare students to apply these tools in physics. Adams et al. (2006) suggested using multimedia to enhance a conceptual understanding of the phenomena

A. Sokolowski, *Understanding Physics Using Mathematical Reasoning*,
https://doi.org/10.1007/978-3-030-80205-9_13

before its mathematizing. Bagno et al. (2008) encouraged discussions of the applicability of the formulas to recognize their broader scientific relationship.

This synthesis illustrates that physics students face difficulties in considering physics quantities building blocks of formulas as covariate parameters. A large scale of attempts to improve that reasoning illustrates that it concerns physics communities. It seems that attempts proposed by research do not aid in seeking more direct links between what students have learned in their math courses but rather at analyzing physics formulas within the context of physics only. This study assesses physics students' awareness of formulas as covariate representations by reflecting on students' mathematical reasoning to uncover that awareness and draws on these findings to propose improvements.

13.2 Theoretical Framework, Research Questions, and Study Logistics

This study, in its initial version, was published earlier (Sokolowski, 2020).

The deduction of inferences in science is possible due to applying covariational reasoning at a larger scale, not just quantifying a physical argument. Mathematical representations that develop students' conceptual knowledge and reasoning skills are algebraic functions in all their forms: symbolic, tabularized, verbal, and graphical (NCTM, 2019). Thus, functions emerge as the foundations of symbolically expressing covariational reasoning. This conclusion guided the formulation of the research questions:

1. *To bring forth the covariational reasoning to the physics classroom, should more parallelism between algebraic functions and physics formulas be enabled to streamline the flow of mathematical reasoning to physics teaching and learning?*
2. *How do students perceive the usefulness of each type of these mathematical constructs: formulas and functions as dynamic functional counterparts to learn physics?*

Thus, rather than further explicating on differences in mathematics and physics perceptions of these two fundamental representations, the pursuit for linking the methodologies of algebra used in mathematics and physics emerged as the study's primary objective.

The study was conducted with a sample of 25 high school students in the US. These students (13 males and 12 females) were enrolled in an advanced physics course after taking an introductory level physics. Eighteen of these students were concurrently taking a precalculus course, and seven were enrolled in an advanced calculus course. While attending the same high school, these students took their prior math and physics courses taught by teachers of various backgrounds and professional preparation before enrolling in the advanced physics course.

The study was conducted over a two-month period that started at the beginning of the semester. Along with traditional applications of formulas to support problem-solving, the students were introduced to physics laws and principles perceived as functional relationships derived either by using an empirical-mathematical model supported by experimentation or by transferring formulas to algebraic functions due to constraints of the scientific context under investigation. At the concluding stage of the study, the students were invited to contrast the effects of using formulas and formulas considered functional relationship on physics understanding.

13.3 Study Instrument

This section summarizes the embodiments of covariational reasoning, its proposed interpretations along with the practices and didactical tasks introduced to the physics curriculum to enhance such reasoning. These methods did not change the scope of the chapters covered, thus kinematics, dynamics, and energy, but they enriched their contents and delivery methods.

The tasks of covariations were clustered into two general domains; (a) applied during experiments and (b) enacted by hypothetical – mathematical covariational models that stemmed from formulas. These structures were integrated within the course of study as they fitted to support the depth of the inquiry of the content under investigation.

13.3.1 General Characteristics of the Treatment: How Did Covariational Reasoning Emerge?

While a broader presentation of covariational reasoning was included in Chap. 5, this part adds specific ideas that were inserted in the traditional physics curriculum and delivered to the students. This section elaborates on the didactical routes taken to develop students' awareness about the opportunities stored in formulas when the formulas are considered as functions. Such formula reimaging was also used to support graphs' sketching and their scientific interpretations. The initial phase of such transitions was brought up by referring to functional representations that students learned in their mathematics courses. Contrasting *side by side* these functions to corresponding formulas was the initial step of the transitions. There were no categories of covariation presented at this time. Such generalization was considered to be deployed when students got familiar with more techniques and have more opportunities to explore the nuances between these categories.

Students study functions of one variable in a typical high school math curriculum, expressed in two different forms: in a coordinate notation, $y = kx$, and the function notation $f(x) = kx$. Functions as means of learning in physics are not emphasized;

instead, physics is populated by formulas perhaps because formulas are concept-specific entities. The formulas are often presented as multivariable expressions, e.g., $F = ma$, which can be converted to a one-variable covariational relationship by considering $F(a) = ma$ or $F(m) = ma$, which is also suggested by Karam et al. (2019). Students realize that the rate of change of each form generates different quantities and graphs in the Cartesian plane, and each, after all, will reflect on a different physical scenario. The nuance of modifying the left side of these formulas has a tremendous effect on the experiment analysis and information learned about its behavior. It was seen that being able to make that recognition was important not only to reflect on the phenomena line of inquiry but also to identify specific covariation embedded in the highlighted quantities. For example, when determining a correct graph of *kinetic energy vs. object's speed* using the formula $K = \frac{mv^2}{2}$, the students first needed to realize that this structure could be sketched considering, for example, the mass, m a constant parameter or a mediating parameter and the object's speed-v a variable parameter, thus $K(v) = \frac{mv^2}{2}$ would reflect the situation and the graph would emerge as a typical quadratic function. Such thinking would lead to realizing that this structure could be considered a quadratic function of the form $f(x) = ax^2$ and sketched as a parabola centered at (0,0) representing the *speed versus kinetic energy graph*. While Karam's suggested deploying this technique to primary functions, this study extended this idea and presented students with techniques of using composite or combinations of functions to support graphs' recognition, thus is targeted scaffolded covariation (as defined in Chap. 5). The following process illustrates the phases of transitioning from primary covariate relations to a scaffolded one. Suppose that an object of mass 10 kg moves with a variable velocity given by $v(t) = 2 + 2t$. Sketch the velocity function for the object for $t \geq 0$. Formulating a kinetic energy function one learns that $K(v(t)) = K(t) = \frac{10(2+2t)^2}{2} = 5(2+2t)^2$. One nuance emerged while analyzing this scaffolded covariation: if $K(v) = K(v(t)) = K(t)$ what independent parameter to label on the horizontal axis; the time, or the velocity? Will the graphs of $K(v)$ and $K(t)$ be identical when either the velocity v or the time t is labeled? While the general shape of kinetic energy function must remain the same, the independent parameter will affect the domain of the kinetic energy function. Two cases emerged:

(a) If time is labeled on the horizontal axes, then the domain of the functions is $t \geq 0$, and the graph will be initiated from a y-intercept labeled in Fig. 13.1 as $K = 20J$. The y-intercept represents the initial kinetic energy of the object with a mass of 10 kg.

(b) If the velocity is labeled on the horizontal axis, then the initial point of the graph will still occur when $t \geq 0$, yet the graph will show up for $v(t) \geq 2\frac{m}{s}$. This graph position in coordinate axes can be supported algebraically, if $t = 0$, $v(t) = 2m/s$, then $K(0) = \frac{10(2)^2}{2} = 20J$, which represents the initial kinetic energy of the object. The graphs are similar in shape but they differ by horizontal transformation (Fig. 13.2).

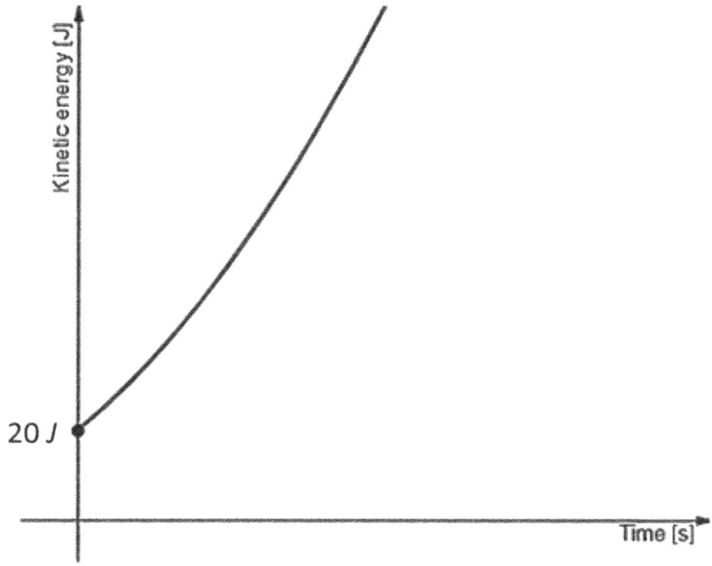

Fig. 13.1 Graph of the kinetic energy versus time of an object moving with a variable velocity $v(t) = 2 + 2t$, when $t \geq 0s$

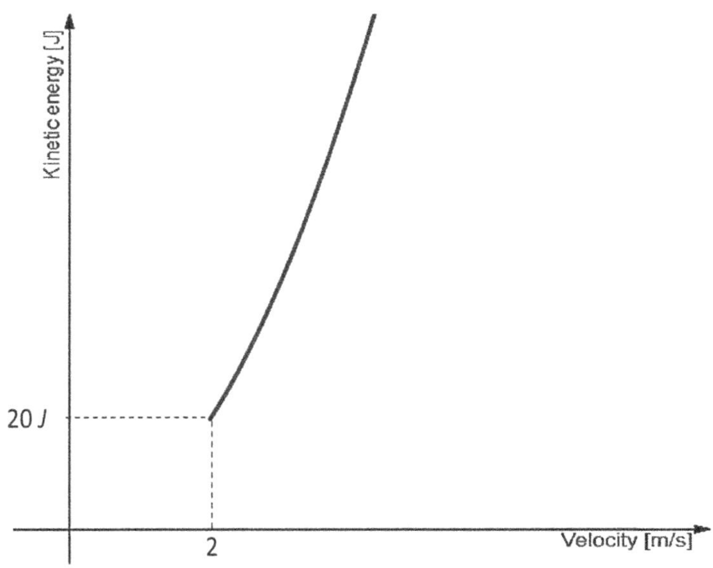

Fig. 13.2 Graph of kinetic energy versus velocity for $v(t) = 2 + 2t$ when $v \geq 2$

Departing from this line of thought, the students were further introduced to sketching composite functions, e.g., $K(v(t))$, where the velocity $v(t)$ was given, for example, as a periodic function $v(t) = Acos(\omega t)$. After creating such composite covariational relation $K(t) = \frac{m(Acos(\omega t))^2}{2}$, the students found it easy to conclude the correct shape for the graphical representation of $K(v(t))$, which was almost impossible without applying the process of formulating explixitly a composite function of the kinetic energy and velocity. Formulas, considered as one covariational relationship, and their composite or scaffold relations helped students conclude the shape of the graph.

Another example of inducing covariational reasoning was the analysis of the formula $T = 2\pi\sqrt{\frac{m}{k}}$. While the formula helps calculate the period of motion of a mass attached to a spring, it comprises of two different objects; m, the mass of the hanging object, and k, the spring constant of the spring, representing two separate independent parameters. In contrast, many physics formulas include parameters related to the properties of one object whose dependent parameter is examined. This duality of the parameters was explicitly realized when considering the formula as a covariational relationship. When examining how each parameter contributed to the period magnitude, it was convenient to split up the analysis in two cases; (a) when one parameter was constant, for example, k and the mass changed, thus examine $T(m) = 2\pi\sqrt{\frac{m}{k}}$, and (b) when the mass was unchanged, and the spring constant was modified, thus examine $T(k) = 2\pi\sqrt{\frac{m}{k}}$. Using $T(m) = 2\pi\sqrt{\frac{m}{k}}$, the students realized that the period increased as the mass increased because as mass increases, the values of the resulting *radical function*, $2\neq\sqrt{\frac{m}{k}}$ also increase. The students suggested making a preliminary graph to support the prediction which was a convenient visual way of verifying the prediction. In (b), the period decreases as the spring value of the constant of the spring increases. This relation could be supported by the behavior of the combination of rational and radical functions. The inner part of the function $\frac{m}{k}$ decreases; thus, $T(k)$ also decreases as k increases. These formulas were further embraced in limiting case analyses. For instance, to prove that the period increases as the mass increases using a limit shows as $T(m \to \infty) = \lim_{m \to \infty} 2\pi\sqrt{\frac{m}{k}} = \infty$ and respectively if (k) is concerned then $T(k \to \infty) = \lim_{k \to \infty} 2\pi\sqrt{\frac{m}{k}} = 0$ that is illustrated in Fig. 13.3.

Both conclusions could be reached using scientific reasoning; however, limiting case analysis conveniently provided an elegant proof. Some students suggested investigating the effects of change of both the mass and the constant spring simultaneously. This case triggered a two-variable covariate that would require a

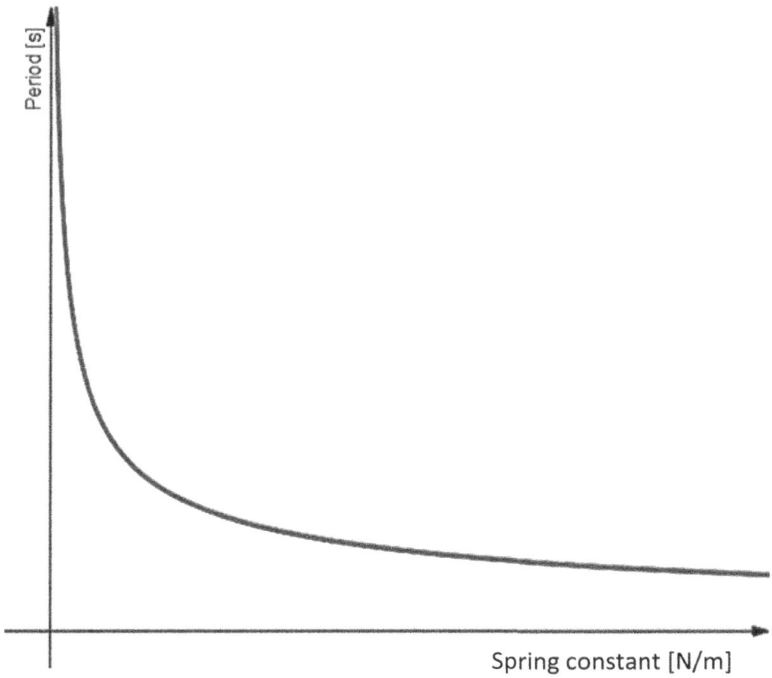

Fig. 13.3 Graph representing the period of a mass-spring system when spring parameter increases

differentiation process. Or such analysis could be supported by experimental data to learn how both the mass and the spring constant affect the period, which perhaps would not be easy to furnish in the realm, but the theoretical analysis could be employed. The graph Fig. 13.3 could also be used to employ the technique of taking sided limit and prove that the period becomes infinite when the spring constant is getting close to zero. This limit analysis created an interesting case to be explained back using scientific terms. Another thought emerged from this analysis; labeling the spring characteristic property as *constant* can, to a certain degree, be problematic. The parameter describing the spring can also be changed by using a different spring or its combinations (series or parallel arrangements), thus, perhaps calling it a mediating parameter or spring parameter, would better describe its property.

Attempts to verify the hypothetical analyzes were made by observing an actual motion of different masses attached to the same spring or the same mass attached to various springs using simulated experiments.

The students realized the limiting case analyzes extended the inquiry far beyond the capacity of the school physics lab. Limiting case analysis was also applied to examine the motion of two blocks on a frictional surface. The methodological details and subsequent evaluation of this method are discussed in Chap. 6. Followed by such inquiries, students' perception of formulas was evolving; the formulas were not considered static entities but as dynamic, continuously changing entities that enrich subsequent inquiries. Students appreciated the opportunities to learn the techniques of considering these formulas as functions because such a view expanded the range of algebraic tools that they could be deployed to support their scientific analyses. What were the areas of weaknesses, and how did physics students perform on questions requiring the induction of more sophisticated math tools? It seems that to answer such questions, more studies are needed. Physics students, perhaps by habits, did not pay much attention to the details of mathematical notation; for instance, the formality of carrying out the process of evaluating function limit. Such perception of mathematical tools did, at times, devaluate the precision and correctness of their answers, although generally, the students developed the correct answers. They also occasionally missed the classification of the parameters as constant or changing, even though experimental setups, verbal descriptions, and general lab conduct provided prompts that explicated these classifications. A general conclusion emerged that developing further students' covariational reasoning skills in physics requires a design of a different curriculum followed by different assessment items emphasising the exploratory character of problem-solving rather than static formula evaluation.

13.3.2 Actions Taken to Exercise Covariation Model Using Laboratory

Mathematical modeling supported by empirical processes was the primary method to encourage covariational reasoning in this study as its effects on developing students' mathematical reasoning in physics are highly recommended. For example, the students observed a bouncing ball and were supposed to derive an exponential decay function representing the amount of mechanical energy dissipated into the thermal energy presented in detail in an earlier study (Sokolowski, 2018). They had an opportunity to apply the limiting case analyses to approximate the total amount of energy lost during the experiment considering the law of conservation of energy as a covariational algebraic entity to initiate the derivation process. Such transitioning, though required the teacher guidance, perhaps due to several covariate parameters that contributed to the formulation of the final function; the students were unsure what parameters to use to build the function. Some included the kinetic energy of the bouncing ball in the general statement, which confused the algebraic process; some did not realize that $Q(n)$; the energy lost by the bouncing ball must be explicitly inserted into the law of conservation of energy during the experiment analysis. This covariate would be classified as Type 4 according to the established scale in Chap. 5. Thus, this experiment is worthy of being discussed in more detail. The initial potential energy of the ball was $U_g = mgh$. The gravitational potential energy retained after each bounce could be denoted as $U(n) = U_g(R)^n$, where $R = \frac{h_n}{h_{n-1}} < 1$ was the ratio of two consecutive bounces. The students concluded that the ratio R was constant by taking data and calculating the ratio; thus,

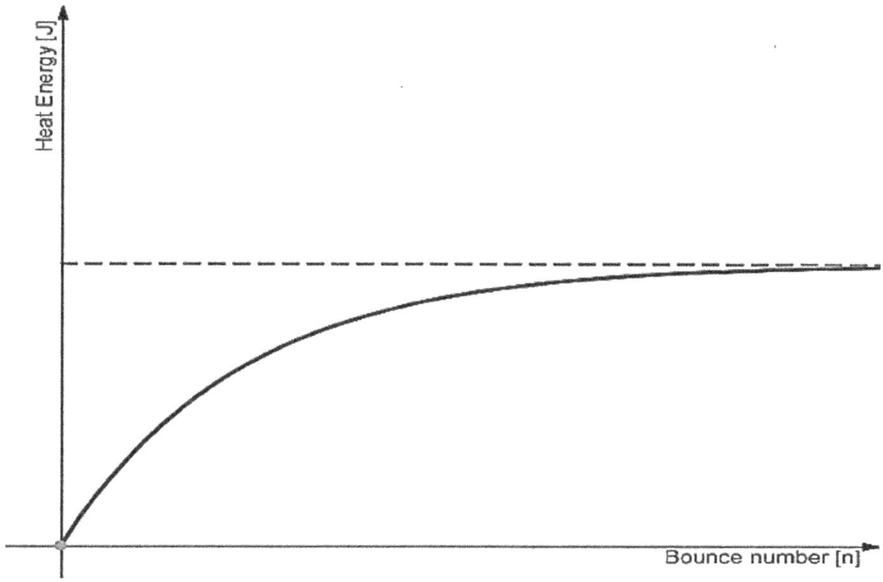

Fig. 13.4 Graph representing dissipation of energy of a bouncing ball

an exponential model could be used. Using the law of conservation of energy resulted in $U_g = U_g(R)^n + Q(n)$. Solving for $Q(n)$ resulted in $Q(n) = U_g - U_g(R)^n = U_G(1 - R^n)$. Since the ratio $R < 1$, the limit $Q(n \rightarrow \infty) = \lim_{n\rightarrow\infty} U_G\left(1 - R^n\right) = U_G$ represents the horizontal asymptote of $Q(n)$ and its graph is lustrated in Fig. 13.4.

The students performed another lab with a congruent context; they analyzed the loss of mechanical energy of a marble oscillating in a vertical parabolic track. After being guided on the *Bouncing Ball* lab, their handling of the lab with an oscilating marble improved.

Conceptual understanding of the cause and effect in experiments is essential; however, identifying the cause and effect and expressing these factors using independent, dependent and mediating parameters is one step toward improving the learning effects from lab conducts. Formulating a covariate expression is the concluding step that prepares the students for more advanced scientific endeavors. Acquiring such skills indeed requires practice, yet the rewards for their possessing are high.

The mathematical models enacted during lab processes are regarded as an axiomatic system that gains significance only after being interpreted through the physical model (Hudson & McIntire, 1977). A meta-analysis (Sokolowski 2015) revealed that modeling generates the largest learning effect sizes compared to traditional teaching methods. Laboratory activities in physics provide excellent opportunities for generating data and expressing the relation between the quantities of interest in a concise language of mathematics. Research shows that most of the graphing tasks in physics focus students' attention on determining the relationship between the quantities of interest without moving the inquiry further and formulating corresponding covariate relations. Without hypothesizing or predicting possible behaviors due to varied conditions, an opportunity to develop mathematical covariational reasoning is lost. Graphing is a crucial aspect of developing a broader perspective on phenomena behavior. The students performed several more lab activities about formulations of algebraic representations of the selected covariational relationships. One of such lab activities used the idea of parametric equations (see Chap. 8) to support the algebraic description of position functions for objects' moving in a two-dimensional plane, simultaneously exercising awareness between position-time graphs and path of motion. Some elements of that lab were explained in more detail in Sokolowski and Capraro (2013). Students were further provided with various types of motion, e.g., projectile, simple harmonic, accelerated on a frictionless plane, and free fall. The task was to establish a frame of reference and determine position functions in two perpendicular dimensions that described the object's motion path. The challenge faced was algebraically expressing the path of motion, for instance, for a free-falling object. The motion path does not represent an algebraic function, yet its symbolic representation was possible to formulate by expressing the height as $0 \leq y(t) \leq y_{MAX}$, and the horizontal location $x = 0$. Another inquiry (see Sokolowski, 2014) had students formulate conditions for speeding and slowing down using a virtual context. Students often associate the state of speeding up with an object's positive acceleration, which is not entirely true when an object possesses a negative velocity. While these states of motion are not signified in the physics curriculum, expressing such modes of movement using a correct sign of acceleration is crucial to formulating Newton's II law of motion. Nevertheless, another purpose of inserting such a lab into the physics curriculum was to eliminate the misconception that an object is speeding up if its acceleration is positive. The students exercised covariational reasoning by aligning the mode of movement of a virtual person with the behavior of acceleration and velocity functions generated by technology. By extracting specific attributes of these covariations, students identified the conditions for speeding up and slowing down and learned how to handle the sign of acceleration when forces and dynamics were concerned.

The prior research documented well students' struggle with graphing. A founding by Planinic et al. (2012) disclosed that physics students do not consider graphs in physics as a means of representing relationships among variables stands for an additional argument to further reimage formulas to functions under favorable conditions. Understanding the scientific underpinnings between quantities of interest was the initial step of formulating a general covariational representation and identifying the attributes of the algebraic expression that fitted that underpinnings concluded unpacking more sophisticated scientific knowledge behind it.

13.4 Data Analysis

After two months of being under treatment described herein, the students were invited to elaborate on the following problem statement:

There are various representations used in physics. For instance, formulas and formulas that can be considered functions.

(a) *Contrast formulas and functions as a means of learning physics.*
(b) *Which of these representations provided you with more opportunities to understand physics concepts?*

Data analysis was initiated by a descriptive assessment of the students' responses clustered in Table 13.1.

Forty percent of the students (N = 10) favored formulas over functions as means supporting the physics understanding. Following Karam et al.'s (2019) classification, these students preferred the technical domain of mathematics. Interestingly, these students (see Table 13.2) highlighted the convenience of applying formulas to solve textbook problems, which is undeniable. Nine students (36%) favored functions, and the remaining six students balanced their responses between both

Table 13.1 Students' preference of algebraic representations in physics

Response Type	Favors Functions Structural Domain	Favors Formulas Technical Domain	Favors Both Formulas and Functions Technical and Structural Domain
Percentage	36% (N = 9)	40% (N = 10)	24% (N = 6)

Table 13.2 Responses favoring formulas

Student	Response
1	In a realistic situation, there will be more than a single variable to plug in to produce an output variable. In physics, there are often many variables. Therefore one must use formulas.
2	Formulas are more valuable to solve problems; functions are used for sketching.
3	Formulas are used to find specific values; functions can change depending on the situation.
4	Formulas are more useful because they have something to do with numbers, while functions deal more with concepts.
5	Functions are essential for creating graphs and visualizing concepts, but formulas create physics.
6	The function is more of a changing thing with many x-values; I like using formulas because formulas are used to plug in specific values to find a realistic result (like $C = 2\pi r$ for the circumference of a circle).
7	I prefer formulas. A function is a set to mathematize a process to produce a value that changes depending on the situation; a formula is the same but useful to find a specific value.
8	Formulas are more straightforward; however, functions allow for more possibilities. Formulas allow changing the value of the variables, whereas functions are often more specific to the data set being used.
9	Formulas are simplified means of finding and explaining an answer in problem-solving. Formulas tend to be my focus, as math is something that more quickly can be researched.
10	Formulas are more frequently used in physics than in mathematics; they are a plug and solve approach without a need for deep thinking.

Table 13.3 Responses favoring functions

Student	Response
1	To solve physics problems, I find functions more useful because functions can be further graphed, and their limits or asymptotes are computed, which supports general comprehension.
2	Functions establish certain constancy when they are graphed. This cannot always be true for formulas.
3	The formula can convey much information regarding the situation, but it cannot display the whole picture and thus only shows instances for the most part. This leads to the strength of a function.
4	Functions can be used to explain concepts in physics on a more intensive level. Learning only formulas can lead to a robotic approach to physics. It can become merely a plug and solve approach without fully understanding what is happening.
5	Functions impart and require more conceptual understanding in their use than formulas do and thus are more useful for genuinely understanding physical concepts.
6	Functions can be much more helpful for certain types of problems than formulas because they allow greater mathematical manipulations such as finding limits, derivatives, and graphing.
7	Functions allow us to visually understand the theory by observing graphs, whereas formulas can be manipulated in various ways to solve for the unknown mathematically.
8	Functions seem to better fit for learning new information.
9	I feel functions are essential resources to utilize when learning concepts in physics because functions can be used in terms of time that allow us to visualize relationships between variables.

functions and formulas. The majority of these students (76%, N = 19) explicitly stated their preference and supported their choices by listing the *pros* and *cons* of applying either function or formula. Samples of students' verbal responses verbatim are clustered in Tables 13.2 and 13.3. While referring to specific answers, T# denotes the table from which the response is cited and S# a specifc student order number from that list. For instance, (T13.2; S4) represents the response of the fourth student from Table 13.2.

A preliminary analysis showed that students consider formulas as algebraic representations that provide a fast way of computing the final answer (T13.2; S6–9). Students highlighted the simplicity of using formulas to solve problems as the primary support for their preferences, which corresponds with conclusions reached in prior research (see Domert et al. 2007). Plugging in given quantities and solving for the unknown is a straightforward algebraic process that dominates current physics assessment items (Uhden et al., 2012). Such an approach leads the students to find solutions and ultimately succeed in problem-solving. This technique is very often exercised, especially in introductory physics courses. The question that arose was if this is the perception of physics, we intend the students to acquire? Even though some students favored formulas, they noted that formulas do not provide means for engaging in a more profound mathematical reasoning to analyze the system's change (T13.2; S4). It was interesting to note that none of these students considered formulas in their original forms as entities that enable graphing but mainly tools to finding solutions to problems. Table 13.3 consists of responses of these students who favored functions.

These students considered functions as entities possessing a high potential of carrying new information about phenomena (e.g., T13.3; S4). A graphing aspect of functions as helpful algebraic tools to understand physics was also prevalent in these responses. There is a visible appreciation of using calculus techniques to derive relations between physics quantities, which originated from students who were concurrently taking a calculus course (T13.3; S6). Students consider formulas converted to functions as a means to support understanding of laws and theorems. Students foresee mathematical reasoning as a gateway to strengthen knowledge of natural phenomena. Formulas are considered as facts or rules constructed using mathematical laws or as physical quantities that are linked with an equal sign. In this regard, formulas appear to students as sets of instructions for creating the desired results. Formulas are viewed as tools to solve the problem, not as tools to explore the system's behavior or learn about a deeper relationship between the parameters of interest. Even when favoring formulas, students perceive functions as superior representations providing more opportunities for exploring relations between quantities. More importantly, they consider functions as covariational relationships generating bases for visualizing these relationships graphically. The students realize that understanding functions require more effort and higher-level thinking skills to encode scientific knowledge efficiently, but such actions' cognitive benefits are high.

13.5 Summary and Conclusions

A traditional perception is that students associate graphing with procedures, not with formulas, perhaps because physics graphs in students' realities do not represent functions at a glance. It was concluded that as long as the students are not explicitly guided on how to view formulas through the prism of covariational relationships, the problem will persist. Graphs in mathematics are generated and classified by their algebraic structures, considering their strict rules that result in curves possessing unique variations. For example, sketching polynomial functions is methodologically different from sketching rational functions. A substantial emphasis on formulas in physics courses makes it almost impossible for students to identify covariational properties and apply them to sketch and analyze graphs for scientific inquiry purposes. This might have another unfavorable impact on physics understanding. If formulas are not converted to functional notations, students might not necessarily perceive sketching curves in physics as sketching equivalent covariate algebraic functions. Considering derived formulas as one variable functions and discussing with students adherence of the resulting graphs to the behavior of the physical phenomena was a technique that seemed to enhance the power of the structural domain of mathematics. Many physics laws and principles have been formulated before a formal definition of function was introduced in mathematics, which perhaps explains the fact that formulas, not functions, dominate the content of physics. Understanding the entanglement of the quantities behind the formulas is an essential component of physics learning because it aids in quantitative descriptions and their underlying theories. Formulas - have been and most likely will - constitute the primary mathematical tool to teach and learn science. However, to follow the recommendations of contemporary research in physics education thus to induce deeper mathematical reasoning to physics education, a shift is needed. Formulas considered functional/covariational algebraic structures to explicate quantities ease the sketching process by making these tasks parallel to what students encounter in their mathematics classes. The techniques of converting formulas to considering these entities as one- variable functions are not often explored in mathematics; therefore, students do not possess the flexibility to switch their perceptions between functions and formulas.

Although the sample size of the participants was small, the general themes that emerged from the data debriefing seem to produce exciting insights on learning physics. One can imply that developing mathematical reasoning in physics can be initiated by enabling covariation reasoning on formulas, which does not require drastic accommodations. The students under treatment realized that many opportunities for deriving new physics knowledge are stored in the functional representation of physics formulas, which was implied from discussions when such conversions were undertaken. It can be inferred from the study outcomes that the problems with knowledge transfer from math to physics lie not necessarily in students' inability to transfer that knowledge but in mathematical representations (formulas) used in physics that in students' realities are not designated to endure such transfer. Formulas are static representations that, in students' perceptions, are used to aid problem-solving rather than to explore system behavior that can, in return, allow for formulating more in-depth scientific theories.

Based on the research findings, it is claimed that teaching students the techniques of converting formulas to one-variable functions smoothens the activations of structural math knowledge, which opens up the gate for using the ideas of rates, limits, concavity, or monotonicity, and graphing. It is claimed that students cannot activate their mathematical knowledge in physics courses because they do not necessarily realize that formulas can be considered as dynamic algebraic representations. It seems that the didactical gap between math knowledge expected from physics students and their ability to handle it can be removed by augmenting and expanding the perceptions on the math tools used in physics. Allowing more functional analysis during physics explorations appears as a task worthy of further research. Undoubtedly, these attempts can not be made at the expense of experimentations that are the core of learning science. However, the study revealed that such attempts enhance the sophistication of students' perception of scientific inquiry moving it to a higher cognitive level.

References

Adams, W. K., Perkins, K. K., Podolefsky, N. S., Dubson, M., Finkelstein, N. D., & Wieman, C. E. (2006). A new instrument for measuring student beliefs about physics and earning physics: The Colorado Learning Attitudes about Science Survey. *Physics Review Special Topics, Physics Education Research, 2*, 010101.

Bagno, E., Berger, H., & Eylon, B. S. (2008). Meeting the challenge of students' understanding of formulae in high-school physics: A learning tool. *Physics Education, 43*(1), 713.

Chi, M. T. (2005). Commonsense conceptions of emergent processes: Why some misconceptions are robust. *The journal of the learning sciences, 14*(2), 161–199.

Cui, L., Rebello, N. S., & Bennett, A. G. (2006, February). College students transfer from calculus to physics. In *AIP Conference Proceedings* (Vol. 818, No. 1, pp. 37–40). AIP.

Domert, D., Airey, J., Linder, C., & Kung, R. L. (2007). An exploration of university physics students' epistemological mindsets towards the understanding of physics equations. *Nordic Studies in Science Education, 3*(1), 15–28.

Hestenes, D. (2003). Oersted medal lecture 2002: Reforming the mathematical language of physics. *American Journal of Physics, 71*(2), 104–121.

Hudson, H. T., & McIntire, W. R. (1977). Correlation between mathematical skills and success in physics. *American Journal of Physics, 413*(13), 470–471.

Karam, R., Uhden, O., & Höttecke, D. (2019). The "math as prerequisite" illusion: Historical considerations and implications for physics teaching. In G. Pospiech, M. Michelini, & B. S. Eylon (Eds.), *Mathematics in physics education*. Springer.

National Council of Teachers of Mathematics (NCTM). An Agenda for Action: Recommendations for School Mathematics of the 1980s. Reston, Va.: NCTM, 1980. Accessed November 13, 2019. http://www.nctm.org/flipbooks/standards/agendaforaction/index.html

Planinic, M., Milin-Sipus, Z., Katic, H., Susac, A., & Ivanjek, L. (2012). Comparison of student understanding of line graph slope in physics and mathematics. *International Journal of Science and Mathematics Education, 10*(6), 1393–1414.

Sherin, B. L. (2001). How students understand physics equations. *Cognition and Instruction, 19*(4), 479–1341.

Sokolowski, A. (2014). Modelling rate for change of speed in calculus proposal of inductive inquiry. *International Journal of Mathematical Education in Science and Technology, 45*(2), 174–189.

Sokolowski, A. (2015). The effects of mathematical modeling on students' achievement; a meta-analysis of research. *Journal of Education, 3*(1), 77–93.

Sokolowski, A. (2018). Modeling with exponential decay function. In *Scientific inquiry in mathematics-theory and practice* (pp. 65–82). Springer.

Sokolowski, A. (2020). Unpacking structural domain of mathematics to aid inquiry in physics: A pilot study. *Physics Education, 56*(1), 015009.

Sokolowski, A., & Capraro, M. M. (2013). A constructivist approach to embodying motion problems in mathematics. *Mediterranean Journal for Research in Mathematics Education, 12*(1–2), 121–133.

Uhden, O., Karam, R., Pietrocola, M., & Pospiech, G. (2012). Modelling mathematical reasoning in physics education. *Science & Education, 21*(4), 485–506.

Teaching Physics Using Mathematical Reasoning

Research and Practice

This book offers novice teaching strategies on applying mathematical reasoning to support the understanding of physics concepts. It proposes an empirical-mathematical scheme to design lectures and lab activities that invite the learners' reason mathematically while attending to physics understanding. The book provides didactical introductions of mathematical ideas that allow physics students to transfer their mathematical knowledge and embrace it in physics inquiries. While there are various levels of mathematical reasoning, this book prioritizes covariational reasoning crafted by limiting case analyses. The volume can interest a broad audience of scholars, practitioners, and curriculum policymakers. It is built on three pillars; (1) synthesis of contemporary research, (2) proposed conceptual framework, and (3) case studies that include readily available lecture ideas that highlight covariational reasoning to support physics understanding. Some of the ideas include; reconstructing Newton's law of universal gravity, parametrization of projectile motion, or enabling covariational reasoning in Einstein's formula for the photoelectric effect.

A. Sokolowski, *Understanding Physics Using Mathematical Reasoning*,
https://doi.org/10.1007/978-3-030-80205-9

Index

A. Sokolowski, *Understanding Physics Using Mathematical Reasoning*,
https://doi.org/10.1007/978-3-030-80205-9

The manufacturer's authorised representative in the EU is Springer Nature Customer Service Centre GmbH, Europaplatz 3, 69115 Heidelberg, Germany. If you have any concerns regarding our products, please contact ProductSafety@springernature.com

Printed and bound by CPI Group (UK) Ltd, Croydon, CR0 4YY

30/09/2026

02225450-0012